Field Guide to the Geology of Northeastern Oman

Geological Field Guides –
Sammlung geologischer Führer

Edited by Peter Rothe

Volume 110

Borntraeger Science Publishers · Stuttgart · 2016

Field Guide to the Geology of Northeastern Oman

Gösta Hoffmann, Martin Meschede, Anne Zacke, Mohammed Al Kindi

With 227 figures

Borntraeger Science Publishers · Stuttgart · 2016

Hoffmann, G., Meschede, M., Zacke, A., Al Kindi, M.: Field Guide to the Geology of Northeastern Oman

Author's addresses:
PD Gösta Hoffmann, Steinmann-Institut für Geologie, Mineralogie und Paläontologie, Rheinische Friedrich-Wilhelms-Universität Bonn, Nussallee 8, 53115 Bonn, Germany.
ghoffman@uni-bonn.de
Prof. Dr. Martin Meschede, Lehrstuhl für Regionale und Strukturgeologie, Friedr.-Ludwig-Jahn-Str. 17A, 17489 Greifswald, Germany. meschede@uni-greifswald.de
Dr. Anne Zacke, Mineralogisches Museum der Universität Bonn, Poppelsdorfer Schloss, Meckenheimer Allee 169, Bonn, Germany. azacke@uni-bonn.de
Dr. Mohammed Al Kindi, Geological Society of Oman, P.O. Box 993 Postal Code: 112 Ruwi, Sultanate of Oman. malkindi@gmail.com

Front cover: "Mother of all outcrops": Deformed Radiolarian Cherts (see EP 24).

We would be pleased to receive your comments on the content of this book:
editors@schweizerbart.de

This publication has been made possible with the generous support by

Geological Society of Oman (GSO)

ISBN 978-3-443-15099-0
ISSN 0343-737X (Geological Field Guides – Sammlung geologischer Führer)

Information on this title: www.borntraeger-cramer.com/9783443150990

© 2016 Gebr. Borntraeger Verlagsbuchhandlung, Stuttgart, Germany

All rights reserved. No part of this publication may be reproduced, stored in a retrieval system, or transmitted, in any form or by any means, electronic, mechanical photocopying, recording, or otherwise, without the prior written permission of Gebr. Borntraeger Science Publishers.

Publisher: Gebr. Borntraeger Verlagsbuchhandlung
    Johannesstraße 3A, 70176 Stuttgart, Germany
    mail@borntraeger-cramer.de www.borntraeger-cramer.de

♾ Printed on permanent paper conforming to ISO 9706-1994

Layout: Satzpunkt Ursula Ewert GmbH, Bayreuth
Printed in Germany by Tutte Druckerei GmbH, Salzweg

Foreword and Acknowledgements

This publication is a joint effort of members of the Geological Society of Oman (GSO) and the German Geological Society (Deutsche Geologische Gesellschaft – Geologische Vereinigung e. V. DGGV). The charter of the DGGV explicitly outlines the purpose of the enterprise to be the facilitation and encouragement of scientific communication to all persons interested in geosciences (§2,3c). The DGGV also aims to foster its relationship with international organisations with similar goals (§2,3k). The main purpose of the GSO is very similar: since its foundation in 2001, the GSO has become a leading society in the preservation of Oman's geological heritage and natural wonders. The society provides seminars, lectures, workshops and field trips to a range of people in various geological fields.

The fascinating geology of Oman is obvious, not only to the expert. Visitors to the country are immediately impressed by the variety of landforms and rocks. For many this comes as a surprise, as the general perception of "Arabia" is often related to dry, barren desert landscapes. Therefore, 3000 m-high mountain ranges or perennial wadis with crystal clear blue water might be unexpected for the first-time visitor. We hope that with this book the reader will be able to put the variety of Oman's beautiful landscape into a scientific context and to understand the processes that led to its formation. Consequently, this book aims to serve as a general introduction and guide to the geology of Oman for laymen, students and geologists alike.

This publication would not have been possible without the support, encouragement and help of many people and organisations. Anne and Gösta explored the country extensively while employed at the German University of Technology. The support of the university management is kindly acknowledged. Often, the field logistics were carried out by Golden Highlands. The support of the company's skillful team was essential in many successful trips, some to very remote areas. Fruitful conversations with geoscience-students from Oman and abroad motivated the compilation of this book.

This work was further supported by The Research Council Oman (TRC) project ORGEBR–10–006 "Short- and long-term environmental changes along the coastline of Oman (Arabian Peninsula)" and TRC project ORG/GUtech/EBR/14/014 "Quaternary sea-level changes in Oman". Furthermore, the help of several individuals is kindly acknowledged: Wilfried Bauer and Wiekert Vissers reviewed the introduction chapters. Thank you to Stefan Schmid for comments on EP 24. Hartmut Seyfried gave valuable comments on EP 19. Adolphe Nicolas and Françoise Boudier shared insider tips on outcrops within the ophiolite. We had productive discussions on boudins and mullions with Wilfried Bauer on EP 98. Wolfgang Frisch proofread the chapter on ophiolites. Paul Yule helped with EP 13. Peter Nievergelt and Dominik Letsch drew our attention to the pillow basalts described at EP 56. Ulrich Kramm shared a lot of his knowledge and commented on EP 30. Comments made by Jean-Pierre Burg helped to improve the description of the complex folds of EP 06 and he also contributed to EP 05, 36 and 37. Rachel Bynoe reviewed chapter 2 which helped to improve the text significantly. The authors are thankful for the professional language review by Stefani Clark. Last not least we also thank Dr. Ibrahim Al Ismaili, the former vice president of GSO, for his support during the early stage of this project.

About the authors:

Gösta Hoffmann graduated with an MSc-equivalent diploma in Geology and Palaeontology from Greifswald University (Germany) in 2000. His research-based PhD project (2001-2004) dealt with the palaeogeographic reconstruction of the postglacial evolution of the southern Baltic Sea coast. He conducted two post-doctorate studies at Utrecht University in The Netherlands and RWTH Aachen University in Germany as a research fellow of the German Research Council (Deutsche Forschungsgemeinschaft DFG). Gösta was among the first staff members in the Department of Applied Geosciences at the German University of Technology in Oman where he joined as an Associate Professor in 2008. He conducted numerous fieldtrips and field classes with GUtech students and led several GSO fieldtrips. Gösta was principal investigator of two research projects, funded by The Research Council (TRC). He also organised and guided expeditions for DGGV in Oman in 2014 and 2015. He received his postdoctoral lecture qualification (Habilitation) from RWTH Aachen University in 2016 and joined the Steinmann Institute at Bonn University as a professor for general geology in the same year. His main research interests are coastal geology and tectonic geomorphology.

Martin Meschede graduated in 1982 from the University of Hannover (Germany) with a diploma in geology. He completed his PhD at the University of Tübingen (Germany) in 1986 and continued his research as a post-doctoral lecturer in Tübingen. Since 2001 he has been a professor of Regional and Structural Geology at the Institute of Geography and Geology, University of Greifswald, Germany. His research interests focus on geodynamic processes at plate margins, subduction, large igneous provinces, exhumation, palaeogeography, palaeoclimatology, basin evolution, and glacial tectonics. He has participated in several marine research expeditions, among them with Joides Resolution of IODP and a diving cruise with the Japanese submersible Shinkai 6500. Besides a number of scientific publications, he is also the author and co-author of several textbooks concerning plate tectonics, structural geology, and regional geology of Germany.

Anne Zacke graduated with an MSc equivalent diploma in Geology at the University of Greifswald in 2003. She gained her PhD in Natural Sciences at the University of Cologne in 2007. Her thesis dealt with palaeoecological and palaeooceanographic reconstructions of warm shelf seas by analysing the apatite of Cretaceous and Paleogene shark teeth. After her PhD she worked as a research fellow at the State Museum of Natural History Karlsruhe in the Department of Museum Education. In 2008, she took the opportunity to move to Muscat and worked as one of the first staff members at the Geoscience Department of the newly established German University of Technology in Oman. She moved back to Germany in 2014 and is currently deputy head of the Mineralogical Museum of Bonn University. Anne works on the interface of science and public, so her main focus is on museum's didactics and education as well as scientific collections.

Mohammed Al Kindi graduated with a BSc in Physics and Geology from Aberdeen University in 2003. Mohammed then obtained a PhD from Leeds University in structural geology in 2006, studying the structural evolution and fracture pattern of the Salakh Arch in the Oman Mountains and relating the surface observations to subsurface data. He worked for Petroleum Development Oman first as a consultant in structural geology, then as a senior production geologist. His main interests include the regional geology of Oman, tectonic evolution of the region and structural setting of hydrocarbon reservoirs. Mohammed has been a member of the committee of the Geological Society of Oman since 2009. He joined the committee as an executive director from 2009 to 2012, then as a president in 2013. He has published and co-authored a number of scientific papers and books about the geology of Oman.

This publication has been made possible with the generous support through the

GSO

The Geological Society of Oman (GSO) is one of the first vocational societies established in Oman. Since its foundation in 2001, the GSO has become one of the leading societies for the preservation of Oman's geological heritage. It provides seminars, lectures, workshops and field trips in various geological fields. Currently, it is involved in designing and establishing a number of geological museums in Oman. There are more than 1,000 members including well-known geological experts. The society also provides geological consultancy and assistance to various ministries in Oman.

DGGV

The German Geological Society (Deutsche Geologische Gesellschaft – Geologische Vereinigung, DGGV) emerged from the fusion of the former German Society of Geosciences (Deutsche Gesellschaft für Geowissenschaften, DGG) and the Geological Association (Geologische Vereinigung, GV) in 2015. The DGG was founded in 1848 as the Deutsche Geologische Gesellschaft and merged in 2004 with the East German Society of Geological Sciences (Gesellschaft für Geologische Wissenschaften, GGW). The GV was founded in 1910. Currently the DGGV has about 3,600 members from universities, industry, and private economy and is the leading member of the umbrella organisation of geosciences (Dachverband Geowissenschaften, DVGeo). The DGGV organises annual meetings, workshops, excursions and is publisher of well-known international journals (International Journal of Earth Sciences, German Journal of Geology).

GUtech

The German University of Technology (GUtech) is a private university in Muscat/Oman. The university was founded in 2007 and the accredited BSc and MSc programs see a growing number of students each year. The vision of the university is to become a leading university of technology in Oman and the wider region, thus defining the highest standards in education, research and innovation. GUtech provides students with the education required to become highly qualified and socially responsible graduates, guided by German excellence in science and technology with a firm grounding in Oman's culture and heritage. The University fosters creative and critical thinking to advance research and development and, through this, aims at serving society as a whole.

Golden Highlands Oman

Golden Highlands Oman is the leading company for scientific fieldtrip logistics and tailor made touristic tours. The company offers individual scientific field campaigns in remote areas as well as logistic support for large field trip parties. See: http://www.goldenhighlands.com/

Table of Contents

Foreword and acknowledgements		VII
1	**Introduction**	1
2	**Archaeology of Oman**	5
2.1	Pre-Islamic Period	5
2.2	Islamic Period	15
3	**Climate**	16
3.1	Modern climate	16
3.2	Quaternary climate changes	17
4	**Vegetation of Oman**	18
4.1	Coast	19
4.2	Desert	20
4.3	Al Hajar Mountains	23
4.4	Dhofar Mountains	25
4.5	Land use	28
5	**Geology of Oman**	30
5.1	Geomorphology	30
5.2	Major tectono-stratigraphic units	34
5.3	Plate tectonic evolution	40
Excursus I: Ophiolites		45
Excursus II: Snowball Earth – the largest glaciation of the Earth's history		52
5.4	Main mineral resources of Oman	57
5.5	History of oil and gas exploration and production in Oman	60
6	**Field sites**	69
References		259
Index list		277

1 Introduction

The Sultanate of Oman is located on the eastern side of the Arabian Peninsula and is well-known for its spectacular geology and rock exposures. For decades, it has attracted numerous geological experts from around the globe who come to study its stunning outcrops and its geological history with a rock record spanning more than 800 million years. They have identified world-class sites that certainly deserve geo-touristic examination. Hence, this book aims to assist people who have an interest in the geology of Oman to find some of these sites and to understand their geological history and context while exploring the wonderful landscape.

The motivation to compile information on geological sites of interest goes back to the establishment of the German University of Technology in Oman (GUtech). Here, as well as at the Sultan Qaboos University (SQU), the education of geoscience students is field-based. The chance to teach geology in the field, where the outcrops act as open laboratories, offers didactical possibilities that go far beyond classroom based teaching. This concept truly enables hands-on education. Of significance is that the field-classrooms are directly accessible, in contrast to areas in other climatic zones where the rocks are hidden behind vegetation and soil cover. Therefore, the field sites in Oman are a global attraction for geoscience students. Not surprisingly, oil companies from different countries use analogue outcrops to train their employees as well.

Another driving force behind this publication is the Geological Society of Oman (GSO). Members of the society utilise these extraordinary outcrops to inform the wider public about the geology in general and to promote the protection of the sites. These activities pave the way to a society who have the knowledge and awareness to protect their natural heritage.

Besides the "Field Guide to the Geology of Oman" by Samir Hanna, published in 1995 (Hanna 1995) and a general geological overview published by Glennie (2005), only unpublished site-specific information existed. The latter are available as field-guide reports made for the fieldtrips of the Geological Society of Oman, which actively runs numerous expert-led fieldtrips every year.

For the purpose of this book, the information from these field guides was compiled in a database. All the sites were visited by the authors and the challenging decision had to be made regarding which sites to include. It was very difficult to exclude any site from the book, as almost all of them seemed worth visiting. We have now chosen 99 sites, which we think are among the most spectacular ones in northern Oman. Our selection is based on uniqueness, petrology and also aesthetic reasons. As a matter of fact, it was impossible to make these decisions impartially, as each author has his or her favourite sites. We would therefore like to encourage the reader to send us information on sites that were not included but worth mentioning in any future edition of this book. As geology is a study of processes, we also know that sites may change due to natural or man-made interventions. Please contact us if any noteworthy changes have occurred.

Oman welcomes a growing number of tourists every year. This is primarily related to its rich culture and fascinating nature including the impressive coast line, majestic mountains and unmissable desert plains. Specialists and laymen are both attracted by the wonderful exposures of rocks in Oman. Therefore, in this guidebook, we aim to:
- Use simplified scientific language,
- enable interested visitors as well as locals to find the sites with the aid of detailed maps,
- increase the awareness to preserve the unique, spectacular and scientifically-important sites,
- provide course material to train local tourist guides,
- promote geo-tourism in Oman.

Our current comprehensive understanding of the geology and landscape of Oman is based on early work by e.g. Miles (1901) or Lees (1928). The area was systematically mapped for the first time by Ken Glennie and co-workers (Glennie et al. 1974) from 1966 to 1968, following the first discoveries of oil and gas reserves. Later on the French geological survey (Bureau de Recherches Géologiques et Minières BRGM), the Geological Survey of Japan and the University of Berne (Switzerland) compiled detailed geological maps in scales of 1:100 000 and 1:250 000. These cover most of northern Oman and parts of the southern Dhofar region (Le Métour et al. 1995) confirming the general structure as outlined by Glennie and co-workers. Detailed accounts of Oman's geological evolution can be found in two volumes published by the British Geological Society: *The Geology and Tectonics of the Oman Region* (Robertson et al. 1990) and *Tectonic evolution of the Oman Mountains* (Rollinson et al. 2014), respectively.

We provide comprehensive descriptions of the routes to reach the described site. This includes detailed and precise maps for the stops. Tremendous improvements have been made to the road infrastructure in the recent years. Therefore, many of the sites that were previously only accessible to four-wheel drive car, may now be visited by normal saloon car. Sites that require an off-road vehicle are highlighted. The investment in infrastructure is ongoing and roads and highways are being constructed literally around the clock. As a consequence, the approaches to the sites as given here may become outdated and we encourage the reader to contact us with any new information. We give the locations as precisely as possible so that the points can be approached using GPS supported navigation tools.

We trust you enjoy the book and wish you many hours of happy exploring!

Code for geological fieldwork, fieldtrips and safety issues

This code helps to raise your awareness and gives some general recommendations on safety precautions when travelling around geological sites in Oman. The text is partly based on "A geological code of conduct" as issued by the Geologists' Association, London. Hence, some of the points discussed below are of general nature and some are specific to Oman.

- Visitors are expected to observe local customs.
- Wear adequate clothing and footwear. Do not wear shorts when going through villages.
- Avoid collecting samples from the sites and remember to "take nothing but photos and leave nothing but footprints".
- Please help to preserve the wild life and do not disturb the natural habitats.
- Do not enter private property without permission of the owner.
- Rock faces are perfectly exposed and in most cases the use of a hammer is not necessary. Please follow the maxim: *mente et malleo,* and think before you hammer indiscriminately.
- Avoid removing in situ fossils, rocks or minerals unless they are genuinely needed for serious studies.
- It is forbidden by law to export rocks, minerals, fossils from Oman, unless you obtain a permission from the Public Authority of Mining.
- Do not remove artefacts from archaeological sites.
- Always keep in mind that you are in a desert environment. Drink and take plenty of water.
- Inform someone of your intended route.
- The weather can change quickly. Wadis (Arabic term for valleys) pose a certain risk as they are prone to flash flooding.
- Please wear light loose clothes, bring a sun hat or veil and use sun block whenever required.
- Make sure your car is in an appropriate condition. Check the tires, including the spare wheel. Make sure you have the necessary equipment to change a tire and make sure you know how to use these equipments.
- Excursion points which can be only accessed by a 4-wheel car are described accordingly.
- Please be careful of snakes and scorpions when you are out in the field. You rarely come across them and they are normally only trying to keep cool or out of your way. Turn rocks over with the foot and wear shoes when doing so.
- GSM network coverage is available in most area. However, we recommend a satellite telephone if you plan to travel in remote areas.
- Sand driving requires special driving skills. Do not drive into the sand desert with one car only. You also need to take a rope and compressor.

- Always remember to wear seat belts and follow speed limits. Driving remains the main safety risk both on blacktop and graded roads.
- It is recommended to burn campfires completely down and not to cover the ash so that the wind can entrain and disperse it.
- Buy firewood or charcoal. Do not chop down trees.
- **In cases of emergency: Call Oman Police / Fire / Ambulance: 9999**

2 Archaeology of Oman

2.1 Pre-Islamic Period

Bronze Age tombs are recognisable landmarks on Oman's mountain ranges, overlooking the countryside, and artefacts from the Stone to Iron Age are common archaeological finds across the Sultanate (Fig. 1).

The oldest evidence for human occupation in the region dates back to the Palaeolithic, which in this region is hypothesised to have begun as early as 1.5 million years (Ma) ago (Petraglia 2003, Cleuziou & Tosi 2007a, Jagher 2009). By this time Oman was populated by hominins who faced the challenge of surviving in an arid environment with high average temperatures throughout

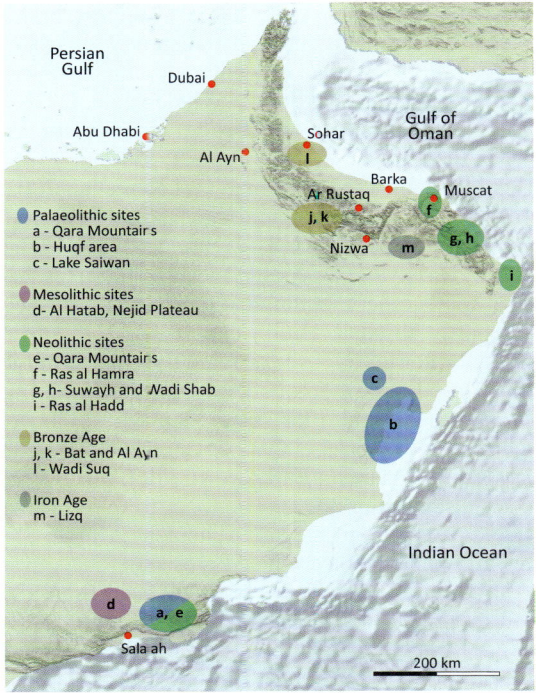

Fig. 1: Some of the most important archaeological sites of Oman (Stone Age to Iron Age) that are described in this chapter.

the seasons and very low rainfall across the country, only with the exception of the highest mountain ranges. This would have, however, fluctuated with changing climates throughout the Pleistocene. Despite these challenges, Oman offers a wide range of different environments suitable for occupation; seasonal rainfall was sufficient to at least fill springs, streams and water ponds in the mountains. Here is where most of the Palaeolithic evidence has been found. The waters of the Indian Ocean are rich in marine life thus providing a bottomless food source for Neolithic societies.

The Stone Age

The earliest archaeological finds are Palaeolithic stone tools that were found within the scope of an archaeological survey in the year 1997 and later on in field campaigns between 2007 and 2009 (Whalen et al. 2002, Jagher 2009). The 1997 campaign focused on wadis in the Jebel Qara/Dhofar district in Southwest-Oman (Fig. 1, Whalen et al. 2002). Choppers, bifaces, scrapers and cleavers, along with their manufacturing flakes were discovered. They were scattered and spread on the surface on gently sloping deflation plains, low-lying areas with occasional seasonal ponds (Whalen et al. 2002, Petraglia 2003). These tools were mainly produced out of chert, which is common in the Paleogene limestones in the region.

Altogether 67 locations were examined (Whalen et al. 2002), most of them on elevated ground overlooking the landscape. Another early Palaeolithic site close to Saiwan/Sharqiyah district was examined in 1984 (Biagi 1994a). The location produced a variety of stone tools such as hand-axes, bi-facial discoids, blade-like flakes and scrapers, again scattered on the surface, covered by a thick patina and partly abraded. Unfortunately, for both sites, neither related stratified deposits nor hominin fossils provide the chance of more exact dating (Petraglia 2003). Based on the frequencies of the different artefacts at each site, the typology present, and after correlation with other locations outside the region, the artefacts of the Jebel Qara have been attributed to the Oldowan and Lower to Middle Acheulean/Early to Middle Pleistocene, about 1.5 Ma (Whalen et al. 2002, Petraglia 2003). The tools from Saiwan are dated on typology as well and date back to the late Acheulean/Middle Pleistocene (Biagi 1994b), but the dating here is uncertain. At present, archaeologists are working on methods to improve the dating methods by establishing an approach for relative chronologies of surface sites.

The campaign from 2007 to 2009 (Central Oman Palaeolithic Survey, COPS) focused on the Huqf region in Central Oman (Fig. 1). 396 locations turned out to be archaeological sites with 340 of them containing tools including cores, light and heavy blades, bi-facials and foliates (Jagher 2009). Acheulean tools, named after the French type site of Saint-Acheul in northern France, are bi-facially worked on both sides and appear to have mainly been used for butchery and woodworking (Cleuziou & Tosi 2007a). The tool-manufacturer was most likely an early ancestor of modern humans, *Homo erectus*. On current evidence, *Homo erectus* appears to be the pioneer among our ancestors who left the African cradle of mankind; the "upright man" evolved approximately 2 Ma ago. Archaeological evidence from the site of Dmanisi in Georgia suggests that their exodus, or at least forays, from Africa into Arabia occurred around 1.8 Ma ago (Gabunia et al. 2000, Lordkipanidze et al. 2006, 2007, 2013). Another finding follows the same line: a *Homo erectus* fragment that was found in Sangiran, Central Java/modern day Indonesia has been dated to 1.5 Ma (Zaim et al. 2011).

The Omani Acheulean sites are the easternmost evidence for Early to Middle Pleistocene settlements in Arabia and contribute to the evidence for a potential transit route out of Africa into South Arabia across the Bab al Mandab Strait that links Africa and Asia. By potentially using a temporary land bridge (Larick & Ciochon 1996) or crossing the open water of the Bab al Mandab (Petraglia et al. 2009) the first hominins may have reached Arabia as early as 1.5 million years ago. The majority of Acheulean sites are likely to relate to post 800,000 year occupations of the region (Petraglia 2003). However, due to the lack of datable material and insecure archaeological findings, the routes as well as timings are difficult to reconstruct.

Once they had arrived, the hominins followed the coast and spread from there along the wadis deep into the interior of the peninsula, where the vast majority of Palaeolithic archaeology is found (Petraglia et al. 2009). The Arabian Peninsula challenged these newcomers with different environmental conditions forcing them to adapt their behaviour for survival. Acheulean finds from Saudi Arabia help to shed light on the lives of the first hominins in Arabia. Their sites were carefully chosen, usually on higher ground, providing high visibility in order to spot possible food resources and fresh water. Additionally, the Palaeolithic settlements are near to lithic resources for stone tool manufacturing. The question of whether or not *Homo erectus* was a hunter or gatherer is still controversial. Possibly they hunted smaller prey but scavenged larger animals and snapped away the prey from competitive predators.

Occupation of Arabia during the Palaeolithic is considered to be discontinuous with only small population densities. The fluctuations in climate which included phases of severe aridity did not favour permanent occupation. As such, populations either shrank considerably or they went extinct during certain periods (Dennell 2003, Petraglia et al. 2009).

Subsequently, the Middle Palaeolithic was a period of major events in terms of human evolution as well as of its dispersal. Around 200,000 years ago, the descendant of *Homo erectus*, *Homo sapiens*, evolved in Africa. *Homo sapiens*

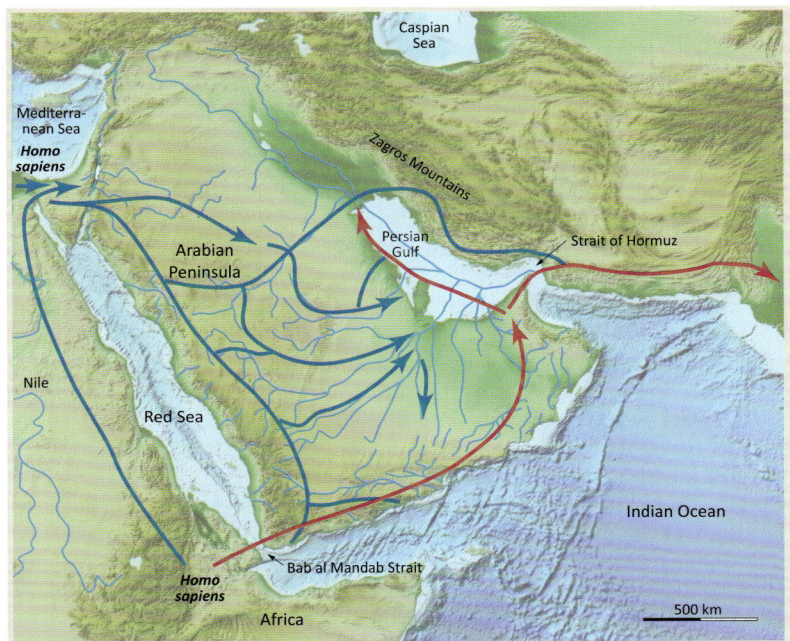

Fig. 2: Modern humans' ways out of Africa: The blue colour marks the northern dispersal route (Nile corridor) and, after reaching Arabia, the spreading by using rivers, lakes and wadis whereas the red arrows indicate the southern dispersal route (Arabian corridor; after Petraglia 2011).

spread, as their ancestors did, out of Africa and into Eurasia but the route and the timing are still uncertain. Two possible routes out of Africa are discussed: migration via a northern, inland dispersal or via a southern, coastal dispersal route. The former would have guided *Homo sapiens* along the western coast of the Red Sea to the Nile River, passing the Sinai into the Levant (Fig. 2; e.g. Vermeersch 2001, Drake et al. 2011). Also named the Nile corridor, this track has been considered to be the primary route for human migration. Within the last few years, evidence has arisen for another, alternative or concurrent route, named the Arabian corridor. This southern dispersal route led from the Horn of Africa across the Red Sea to southern Arabia (Fig. 2; e.g. Lahr & Foley 1994, Stringer 2000, Rose & Petraglia 2009, Armitage 2011).

Whatever route *Homo sapiens* took, the timing was strongly linked to the prevailing climate. Arabia's vast dune fields build a natural barrier and were considered to be major obstacles for the movement of humans. Archaeological finds from Saudi Arabia and the United Arab Emirates indicate that migration patterns of early humans were strongly linked to pluvial phases of interglacial periods of mild weather between the ice ages that facilitated the crossing of the Arabian Peninsula. *Homo sapiens* could have used an early climatic window that opened between 135,000 and 120,000 years ago (late Marine Isotope Stage (MIS) 6 to 5e) or a later window between 82,000 and 78,000 years ago (MIS 5a, Armitage et al. 2011). Importantly, interglacial periods were also times of high sea level, which had implications for the crossing of the Bab al Mandab strait. *Homo sapiens* obviously managed to cross this obstacle; the fascinating question of how they did it, remains open.

The discovery of Mid-Palaeolithic sites – again in the Dhofar region in the south of Oman (Fig. 1) – during an archaeological campaign in 2010 and 2011 produced a breakthrough with new answers. Hundreds of tools, points, blades, side-scrapers and perforators, dated to an age of 106,000 years, show close affinities to tools discovered at localities in Northwest Africa called the Nubian complex. The observed similarities are seen as a result of cultural exchange between the Arabian and African populations (Rose et al. 2011, Usik et al. 2012). The findings are "… stone breadcrumbs – spread across the southern dispersal route out of Africa." (Rose et al. 2011: p. 18).

The fate of the Nubian toolmakers remains unclear. Neither their exact time of arrival nor their disappearance could be reconstructed. The prevailing climate in the region changed after MIS 5 (between 80,000–130,000 years ago) to extreme arid conditions, but evidence from terrestrial archives constrains wetter periods during early MIS 3 (between 60,000–50,000 years ago), probably allowing a demographic north-south exchange (Rose et al. 2011).

Evidence from the Mesolithic is rare in Oman, as sea level rise is likely to have obscured much of the coastal evidence. A few sites have recently been discovered but the outcomes of the excavations are not published yet. There is one site, Al Hatab, which is dated to the Late Palaeolithic to Mesolithic (Rose & Usik 2009). The site was mapped during the field campaign of the Central Oman Pleistocene Research Program from 2002 to 2008. Al Hatab is a partially collapsed rock overhang located at the Nejd Plateau/Central Oman (Fig. 1). Here, nearly 2000 chopped stone artefacts were found (Rose & Usik 2009). The conclusions drawn from this campaign state that the South Arabian Late Palaeolithic culture "… probably belongs to a unique and locally-derived lithic tradition" (Rose & Usik 2009: p. 182), that might have been loosely related to the Levantine.

The transition to the Neolithic is difficult to define as the rare surface sites do not produce reliable findings for dating nor evidence for the origin of the tools. In the caves and rock shelters of Jebel Qara, a limestone massif in the Dhofar region (Fig. 1), the oldest remains of Holocene hunter-gatherer societies were found. These remains included debitage, all the different material produced during the making of chipped stone tools, as well as Fasad points, named from the Ramlat Fasad in Dhofar where they were found for the first time. They are associated with high concentrations of land snails, lenses of charcoal and fragments of burned bones. The accumulation of the molluscs is interpreted as anthropogenic accumulation, meaning the collection was formed by humans who were consuming them (Cremaschi et al. 2015). Their presence in such vast quantities provide evidence of the foraging strategies of human societies between 10,500 and 8,000 cal BP, with a climate optimum culminating at around 9,000 to 8,000 cal BP (Cremaschi et al. 2015). The Holocene societies exploited the snails as a food resource. The inland occupation of this area was only possible due to a climatic shift to wet conditions. Around 9,000 cal BP the climate shifted back to arid conditions and Jebel Qara was abandoned. The Neolithic groups of hunter-gatherers appear to have moved to coastal regions with more favourable living conditions, including access to marine resources.

It remains unclear whether or not Arabia was continuously populated by hunter-gatherer societies or became abandoned while unfavourable climatic conditions prevailed. The present archaeological picture is too fragmented, with considerable chronological and spatial gaps, to answer this question. In addition, at certain points on the coast there is the possibility of earlier evidence being submerged.

Younger Neolithic finds are scattered along the coast (Fig. 1). Archaeologists are able to define distinct Neolithic technologies covering a time span from the already named Fasad-technology to an industry that is dominated by tri-hedral arrowheads (6,500 to 4,500 Before common/current era (BCE); Méry & Charpentier 2013) and another characterised by fusiform foliate artefacts (4,500 to 3,700 BCE). The societies along the coast used different fishing techniques such as net, line fishing and fish traps with domestic animals permitting a certain independence from the availability of prey seen at sites such as Suwayh (e.g. Charpentier 2008), Wadi Shab (Tosi & Usai 2003) and Ras al Hamra (Salvatori 2007). Shell middens along with rich cultural material are often found along the coast and indicate that Neolithic cultures exploited the rich mollusc fauna as well (Charpentier 2008).

In this time frame, in neighbouring regions like Mesopotamia, the first states were already developing. But the societies living on the Arabian Peninsula persisted in a "Stone Age subsistence economy" (Cleuziou & Tosi 2007b) with small settlements without specialisation. The role of agriculture remained marginal. Many of these settlements, dated around 4,000 BCE, have been discovered, for example at Ras al Hamra in Muscat, Quriyat, Tiwi, Wadi Shab, Sur, Ras al Hadd and Ras al Jinz. At that time, local groups consisting out of a few families, mostly four to ten households with altogether thirty to sixty people, moved between seasonal campsites (Cleuziou & Tosi 2007b). Archaeological findings at the sites include ornaments like beads and earrings, tools like shell-hooks, net sinkers and stone tools but also some pottery imported from central Mesopotamia. Although known in the communities for a fairly long time, probably at least by the end of 6,000 BCE (Cleuziou & Tosi 2007b), they did not manufacture products like pottery themselves.

The youngest Neolithic tools are dated to around 3,100 cal. BCE. These are mainly end- and side-scrapers with a large spatial distribution leading to the conclusion that a trade-network might have been in use (Charpentier 2008). It was around this time that the transition from subsistence economy to a productive economy happened.

Copper and Bronze Age

Around 3,100 BCE, farming became established and technologies such as pottery production and metalwork were introduced. The source for the latter was right on their doorstep: the rocks of the Samail Ophiolite contain rich copper deposits. Once these deposits were discovered and the value of these ores recognised, Oman was immediately incorporated in a maritime exchange network. The trade routes connected the Arabian Peninsula with Mesopotamia, Central Asia and the Horn of Africa by sea and overland.

Some of the most striking prehistoric monuments in Oman also date from the Bronze Age. Cairn graves – also called beehive tombs due to their distinctive shape – are visible all over Oman, often located along mountain ridges and thus visible from long distances. Estimations place more than 100,000 of these tombs across the Sultanate (Cleziou & Tosi 2007c), with groups of dozens to hundreds in one area being common. The name 'cairn' was given by a Danish archaeologist in the 1960s. It comes from Scottish Gaelic and means 'a pile of stones' which were often used as way markers. Overall, these were erected between 3,300 BCE and 1,000 BCE (Cleziou & Tosi 2007c).

The oldest graves are called 'Hafit' cairns (3,300–2,700 BCE), which form the majority found in Oman with some of the most spectacular ones found at the necropolis of Al Ayn (Fig. 1; Cleziou & Tosi 2007c). Hafit tombs are conical-shaped circular towers with one chamber; the entrance faces east. Building materials are either worked stone slabs or unworked stones, all from the surrounding geology. The towers, up to eight meters high, were piled-up with two to three walls encasing the burial chamber. Most of the graves were plundered but some remain untouched, containing rich finds like buried skeletons and grave goods such as beads, copper ornaments and needles (Cleziou & Tosi 2007c). The graves were probably used over generations for collective family burials, as many of them contain more than one skeleton.

The 'Hafit' type was replaced by the 'Umm an Naar' type around 2,700 BCE, which were in use until approximately 2,000 BCE. Often built on plains near settlements, they became damaged over time and are less eye-catching nowadays than their precursors. 'Umm an Naar' tombs were built with a larger diameter but are lower in height, around three meters. The inner structure was more complex; several chambers provided more space for more burials. Two entrances allowed people to enter the graves. About 100 'Umm an Naar' tombs have been discovered. Compared to the 'Hafit' type, the smaller number of graves might be explained by the re-use of the carefully worked stones of these tombs over time although we see re-use with the Hafit type too. In Bat and in Al Khutm (Fig. 1), another necropolis was discovered (e.g. Cleziou & Tosi 2007c). Here both types of graves are common. Along mountain crests and the foothills to the north of the oasis, a large number of 'Hafit' type tombs were erected, whereas the younger 'Umm an Naar' tombs cover the plain. Al Ayn, Bat and Al Khutm are protected by the UNESCO convention. The organisation states that this "… is the most complete and best known archaeological complex in Eastern Arabia for the 3rd millennium BC" (UNESCO 2014a). In general, the graves were used over generations for family members of both genders and all ages.

At the time when the tombs were built, Oman was known as the country of Magan. Sumerian (Magan) and Akkadian (Makan) cuneiform texts dated from 2,500 to 1,800 BCE report about this centre of copper extraction and smelting (Weisgerber 2007a). The products were exported to Mesopotamia and probably as far as the Indus Valley (Weisgerber 2007a). The people used stone hammers and additional metal chisels to break the copper-bearing ore. Afterwards, the ore was crushed. The ore was then smelted by burning charcoal over temperatures of more than 1100 °C. In small pear-shaped and knee-high furnaces out of clay (Weisgerber 2007b). In order to reach such high temperatures, the furnaces had holes for ventilation and bellows were probably used to aid the smelting-process. The fluid metal was then poured into holes where the metal took its typical bun-shaped form (Weisgerber 2007b). From the spatial distribution of places of copper smelting, it appears that the precious metal was transported to the western shores of Oman and shipped from there to its final destination.

The people in the Bronze Age probably used the hot summer months to exploit the copper mines in the mountains, where temperatures were lower, moving to their coastal settlements during the winter months, when the rich marine resources could be fully exploited. Contemporary renovation work on buildings indicates that the houses were left for some time and restored on their return (Logers 2015). It seems clear that the people of the country of Magan lived half-nomadic lives in almost egalitarian clans without establishing more complex societies. The long-distance trade economy brought higher living standards and social changes to the societies; the complex of Bat indicates a more hierarchical and structured social organisation (UNESCO 2014a).

One of the most remarkable rock carvings – the "Hasat Bin Salt" group of several persons – date probably from the Bronze Age, too (see EP 69). The group consists of at least seven people: males, females and one child. Style and size of the figures are nearly unique, however, the age as well as purpose and meaning of this rock carving remains controversial.

With its remarkable settlements and necropolises, the newly adopted technologies like pottery production and metalworking, the early Bronze Age civilisation had obviously reached its zenith – with a certain wealth and higher living standards than before (Cleziou & Tosi 2007d).

Findings from Wadi Suq, west of Sohar (Fig. 1), dated to 2,000 BCE, tell the story of the subsequent decline of the early Arabian civilisations (e.g. Cleuziou & Tosi 2007e). The so-called Wadi Suq period (2,000 to 1,300 BCE, e.g., Cleuziou & Tosi 2007e) can be distinguished from the older Bronze Age cultures by a new type of pottery and is marked by impoverishment and depopulation. It seems that the Magan society collapsed probably within no more than a century and became replaced by a new social order (e.g., Cleuziou & Tosi 2007e). Climate fluctuations as well as the collapse of the late third millennium trade are discussed as reasons for this development (e.g., Cleuziou & Tosi 2007e). Sites of Wadi Suq age have been located all over Oman.

The Iron Age

Although this period is called the Iron Age, the metal iron came into general use in Oman only in 400 BCE, almost at the end of the Iron Age. However, around 1,300 BCE with the end of the Wadi Suq period, transformation started resulting in a new regional culture that was finally established all over Oman around 1,000 BCE. The pottery of the earliest stage of the Iron Age, Iron Age I (1,300 to 1,100 BCE), shows only a few distinctive characteristics thus making it difficult to recognise. Additionally, it was not given as a luxury item to the dead anymore. At this time, small communities lived in fortified settlements whereas other groups preferred a nomadic life (Cleuziou & Tosi 2007f). One notable site of this age is located close to Nizwa in central Oman (Fig. 1). In the ruins of a remarkable hillfort (named Sharjah al Hadirah) weapons like a battle-axe, daggers, arrowheads etc. were found. The place was interpreted as the grave of a wealthy warrior (Cleuziou & Tosi 2007f).

Archaeological evidence from the Iron Age II (Lizq period, 1,100 to 600 BCE) is less rare and more significant. In general, this time is characterised by a remarkable cultural unity (Cleuziou & Tosi 2007f). Almost all abandoned oases became resettled, new falaj systems were built, and the pottery of this period was painted again. The population grew – Iron Age II settlements are known from all ecological compartments: the plains in front of the mountains, the coast, the wadis, the mountains. The settlements included villages, fortified villages or hillforts out of mud-bricks or stones as well as campsites. Outstanding sites of this age have been discovered in the UAE. The houses, which were excavated there, had only small windows just below the roof as an adaption to the hot climate. On average three to four rooms, plastered floors and elevated sills against sand and animals entering the houses are additional characteristics (Cleuziou & Tosi 2007f). Similar villages are known in Oman as well, for example north of Ibri or in the Arja copper mining area west of Sohar (Cleuziou & Tosi 2007f). All these villages were associated with a falaj.

In Lizq (Fig. 1), the best known hillfort from this period, named Lizq period, was discovered. The fort was built by using stone blocks fixed with mortar. In general, these hillforts, sometimes overlooking the villages, indicate a certain complexity of social and political life of Iron Age II societies. Agriculture was widely developed with palm trees, fruits, cereals and legumes. Fishing provided another source of food. The camel became domesticated; there are ongoing discussions about the place and timing of this process. However, the animals were important for the nomadic part of the population as they enabled them to move further into the desert. Additionally, nomads could move faster and in larger groups. The use of the camel as pack animal for land trade was another important development (Cleuziou & Tosi 2007f).

In the villages, the communities were probably still kin-based, organised with leading families in the forts and extended families in the houses next door (Cleuziou & Tosi 2007f). Their graves, mostly for individual burials, show a high variability in the form of cairns and also a variety of shapes: round, rectangular and oval. The rich archaeological evidence from this period also includes abundant pottery, copper weapons, soft stone vessels as well as beads (Cleu-ziou & Tosi 2007f).

The production of copper restarted on a large scale. Smelting took place near the ore body. Remains of a furnace were only found at Wadi Qatif. Near Yanqul a settlement was discovered, where copper was produced over a long period, from 1200 to 800 BCE (Weisgerber 2007c). Along the Omani coast, several sites of the Iron Age were identified as well. The people here still exploited marine resources as indicated by shell-middens, which were found for example in the Ras al Hadd area (Cleuziou & Tosi 2007f).

After 600 BCE, a transformation seemed to be happening as new types of pottery appeared that showed strong similarities to Iranian products. At this time, the region was dominated by Persian rulers of the Achemenian Empire or First Persian Empire (550–330 BCE). After the fall of the Persian rule different dynasties influenced the region. It became a province first of the Parthian (100 BC–250 AD) followed by the Sasanian empire (250–650 AD, called

Samad period). The Romans, as counterpart to these empires, only controlled rich trading tribes in northern Arabia. The archaeological evidence of this time argues for continuity: earlier houses were reoccupied; new houses were built in the style of the old ones. On the other hand, trade brought new goods into the region and the material culture changed deeply. Iron was now the dominant metal; the use of coins was introduced by the Greek (Cleuziou & Tosi 2007f). In the 7th century, during the lifetime of the prophet Muhammad, the people in the region adopted Islam and a new chapter in Oman's history started.

2.2 Islamic Period

There are different versions how the Islam spread over Oman. However, the different sources agree that Muhammad the prophet has sent letters to the rulers of Oman convincing them to convert to the Islam and to encourage others to follow their example.

After the letters found their addressees, the religion spread quickly during the 7th century in the Arabian Peninsula. The archaeological evidence from the early Islamic Ages is sparse. Huge amounts of slag deposits indicate that copper mining and smelting took place where the upper part of the ophiolite sequence was easily exploitable. Weisgerber (1991) assumes that the copper production was at maximum during this period. Other archaeological evidence is difficult to find, the oldest buildings in Oman date back to medieval ages.

Over the following centuries Oman's changeful history became written down in documents and books. Brief summaries can be gleaned in every guidebook.

3 Climate

3.1 Modern climate

At present, most of the Arabian Peninsula is dominated by desert environments. The climate is semi-arid to arid and locally even hyper-arid (Glennie & Singhvi 2002). With some local exceptions, most parts of southern Arabia receive between 50 and 200 mm of annual precipitation. Precipitation events are sporadic, highly variable in space and time, and can be intense with a high erosional potential. The quantity and duration of precipitation is irregular.

The principal mechanisms responsible for precipitation are summarised by Kwarteng et al. (2009) as convective rainstorms, tropical cyclones, cold frontal troughs and the southwesterly monsoon. The latter is of importance in the southernmost part of the peninsula, namely parts of Yemen and the Dhofar region in southern Oman. Here, considerable amounts of precipitation are received throughout the summer months which are associated with the northward movement of the Intertropical Convergence Zone (ITCZ) towards the southern part of the Arabian Peninsula. The local name for the rainy season is the *Khareef*. Northern and central Oman receive the most rainfall during February and March. Here, the Oman Mountains act as a natural barrier and cause an orographic effect, a shift in atmospheric condition caused by a change in elevation, resulting in convective rainfall.

Oman is subject to infrequent tropical cyclone influence. The storms that develop in the Northern Indian Ocean only rarely affect Arabia, as they either dissipate or turn towards India (Murty & El-Sabh 1984, Blount et al. 2010, Fritz et al. 2010). Oman's vulnerability to such extreme events became apparent after the landfall of Cyclone *Gonu* in 2007, Cyclone *Phet* in 2010 (Hoffmann & Reicherter 2014) and Cyclone *Chapala* in 2015. Cyclone *Gonu* is the most intense cyclone on record in the Arabian Sea. The total amount of rainfall (~600 mm) exceeded the average yearly rainfall tenfold.

Apart from monsoon-driven precipitation, the climate system of the Arabian Peninsula is also influenced by air masses originating in the Mediterranean in the form of mid-latitude Westerlies (Enzel et al. 2008), especially during the winter time. Northwest to southeast blowing low-level winds that occur during the summer months are locally known as the *shamal* (see Glennie & Singhvi 2002, Parton et al. 2015).

The temperatures show strong seasonal variations in Oman. Whereas the average temperature in the summer-months ranges between 32 °C and 48 °C, with local maximum temperatures exceeding 50 °C, the temperatures are considerably lower from October to April (Kwarteng et al. 2009). Hence, the winter months are the ideal period for fieldtrips with mean temperatures ranging from 26 °C to 36 °C. The highest peaks of Oman which are 3006 m high, occasionally record temperatures around zero degrees Celsius during January–February. In the hot summer months humidity in the coastal regions may rise to 90 % and even higher.

3.2 Quaternary climate changes

The climate of the Arabian Peninsula shows considerable variation throughout the Quaternary (see EP 32, 48, 52, 53), resulting in drastic arid-humid transitions and fundamental environmental deviations, which had an important impact on past human societies (Parker & Rose 2008, Petraglia et al. 2015). The Quaternary climate variability is summarised by Hoffmann et al (2015). The general conclusion is that the Quaternary climate history of southern Arabia was characterised by changes in precipitation, where the humid periods primarily coincided with interglacial periods at higher latitudes, whereas glacial phases were rather arid (e.g., Weyhenmeyer et al. 2000, Fleitmann & Matter 2009). Although, the summer monsoon activity was weaker during glacial periods, the western convection was probably stronger (Rohling et al. 2013).

Climatic conditions are also of high relevance for the potential hominin dispersal across the southern Arabian Peninsula, which is expected to have acted as a bridge during pluvials and a barrier during arid phases (Parker & Rose 2008, Parker 2009, Rosenberg et al. 2011, Parton et al. 2015). A pluvial is a period of time spanning from a decade to thousands of years that are very wet or particularly humid. Archaeological evidence indicates the existence of human hunter/gatherer populations in the north of the Oman Mountains during MIS 3 (Armitage et al. 2011).

Quantitative determination indicates that precipitation during the wetter periods was up to five times higher than present (Woods & Imes 1995). The climate variability recorded in terrestrial archives of southern Arabia, notably the change in moisture supply, is apparently controlled by changes in the location of the ITCZ and the associated rainfall belt of the Indian Ocean Monsoon (IOM). During times of increased summer season insolation the mean latitudinal position of the summer ITCZ shifts northwards and the IOM is drawn into the continent. This results in a rainy season which also affects northern Oman (Fleitmann et al. 2003). Periods that witnessed a strengthening of monsoon circulation and increased humidity within the interior of the Arabian Peninsula are recorded for the Early to Middle Holocene. Some phases during the Pleistocene were also wet, whereas massive aeolian deposition implies that also periods, characterised by arid to hyper-arid conditions existed (e.g., Radies et al. 2004).

4 Vegetation of Oman

Author: PD Dr. Peter König
Affiliation: Ernst-Moritz-Arndt-Universität Greifswald; Botanical Garden; Soldmannstr. 15; D-17487 Greifswald, Germany

The main ecological regions of the Sultanate of Oman are built up by the Al Hajar and Dhofar mountains with extensive deserts, i.e. the Jiddat al Harasis, the Rub al Khali and the Wahiba Sands, in between. The Gulf of Oman and the Arabian Sea encircle the country and form the long coast line (Fig. 3). About 1200 plant species form the basis of the country's vegetation (Ghazanfar 1999, Pickering & Patzelt 2008).

Fig. 3: Landscapes of Oman, top left clockwise: a) *Acacia* woodland of the foothills, b) Musandam mountains, c) *Juniperus* woodland in the Hajar highlands, d) sand dune vegetation in the Wahiba Sands, e) Reg area in the Jiddat al Harasis bare of vegetation, f) monsoon forest in Dhofar, g) open dwarf shrub vegetation in the central desert.

4.1 Coast

The extensive sea shore of Oman is home to a halophytic vegetation with a clear-cut zonation. Halophytic plantlife thrive in salt water environments. At sheltered places in lagoons (Arabic Khawrs), the Grey Mangrove (*Avicennia marina*; Fig. 4a) manages to survive in sea water. Pneumatophores or respiratory roots help the plants to satisfy the demand for oxygen during low tide. After being overexploited or damaged by cyclones, e.g. Cyclone Gonu in 2007 or Cyclone Phet in 2010, recently, numerous efforts in reestablishment and reforestation have been made.

Coastal mud flats and dunes harbour a broad set of specialists like *Arthrocnemum macrostachyum* (Salt bush family; Fig. 4b) or *Sphaerocoma aucheri* (Pink family), and *Aelurops lagopoides, Sporobolus spicatus, Urochondra setulosa* (Fig. 4c) among the grasses.

Fig. 4: a) *Avicennia marina* during low tide, pneumatophores exposed. b) Flowering *Arthrocnemum macrostachyum*, a salt resistant succulent. c) *Urochondra setulosa* on coastal dunes.

4.2 Desert

Salt pans or *Sabkha*s are bare of any vegetation in their central parts. The Umm as Samim ranks among the most extensive of the inland *Sabkha*s in Oman. In the vicinity, plant life starts at micro-dunes, so-called *Nabkha*s, small patches, where first sand mounds evolved, enabling plant and root growth above the

Fig. 5: a) The "String-of-beads" (*Halopeplis perfoliata*) turns deeply red when salt concentration is at the limit. b) *Tetraena qatarensis* is one of the frequently seen desert plants. c) The saltwort *Salsola rubescens* is densely covered by a hairy indumentum, a covering of fine bristles, and carries winged fruits (orange). d) *Heliotropium bacciferum* is a prickly perennial with fleshy fruits. e) The parasite *Cistanche phelyphaea* lacks any green and hosts on various shrubs, the flower colour ranges from yellow to purple tinged.

deadly salt level of the pan surface (König 2012). *Sabkha*s are also common along the coast line where sea water inundates depressions. The "String-of-beads" (*Halopeplis perfoliata*; Fig. 5a) is a typical succulent dwarf-shrub seen in these locations and a circum-Arabian character species, found on nearly all coasts along the Persian Gulf and the Red Sea.

Extensive gravel deserts are covered by open dwarf-shrub communities (Fig. 3g). The "pair of leaflets" plant *Tetraena qatarensis* (Fig. 5b) is most successful in facing the harsh arid conditions. Leaf quantity and structure (petioles, unifoliate or bifoliate leaves) are controlled by the moisture status of the environment, thus enabling the species a high flexibility. This succulent is not sensitive to grazing, because it is mostly avoided by livestock.

The heliotrope *Heliotropium bacciferum* (Fig. 5d) and the chenopods *Halothamnus bottae, Haloxylon salicornicum, Salsola rubescens* (Fig. 5c) rank among other more common desert plants.

Huge areas covered by inland dunes are found in the Wahiba Sands (Fig. 3d) and the Rub al Khali, the latter part of the "Empty Quarter", making up the bulk of the southern Arabian Peninsula. Species well adapted to sand movement are the sedge *Cyperus conglomeratus* (Fig. 6a), Riebeck's Spurge (*Euphorbia riebeckii*), which is endemic to Oman, or *Calligonum comosum*, a member of the Knotweed family. Ghaf trees (*Prosopis cineraria* Fig. 6b) are quite common and form woodlands in some parts of the sand deserts. Their deep-reaching root systems enable the plants to tap into the underground water storage filled by the wide-pitted capillary system of the sands.

Wadis, i.e. dry river beds outside a rain event, are common. Their size and vegetation differ with the local site conditions. In the mountains, water flow velocity is high; wadi beds consist of gravel and big boulders, and the Willow-leaved Fig (*Ficus cordata* subsp. *salicifolia*) may grow in crevices. With decelerating flow, the gravel is more fine-grained, and site conditions give place to *Pteropyrum scoparium* (Fig. 6c), a shrubby knotweed with winged fruits endemic to Oman and UAE, or the Christ Thorn (*Ziziphus spina-christi*; Fig. 6d). The latter is one of the "Bible plants", the epithet *spina-christi* refers to the crown of thorns worn by Christ at the crucifixion.

The lower wadi courses stretch far into the desert and mainly Acacias like the Umbrella Thorn (*Acacia tortilis*; Fig. 6e) or Ehrenberg's Thorn (*A. ehrenbergiana*) feed their roots from this source and dominate the vegetation.

Fruits of the "Desert Pumpkin" (*Citrullus colocynthis*; Fig. 6f) are commonly seen along extensive wadi beds. The fresh fruits are used for their purgative properties; the dry fruits break off from the parent plant and roll over the ground in strong winds or are washed away by flash floods, later breaking open and scattering their seeds.

Fig. 6: a) *Cyperus conglomeratus* is a typical plant of the sand dunes. b) *Prosopis cineraria* woodland in the Wahiba Sands. c) The knotweed *Pteropyrum scoparium* is a typical wadi plant of the northern mountains. d) Fruits and leaves of the Christ Thorn (*Ziziphus spina-christi*) are sold at the markets as a sweet and a shampoo respectively. e) The Umbrella Thorn (*Acacia tortillis*) forms xeromorphic woodlands. f) Ripe fruits of the „Desert Pumpkin" (*Citrullus colocynthis*) are orange-sized.

4.3 Al Hajar Mountains

Numerous wadis run down the northern escarpment and slopes, forming huge alluvial gravel fans along the northern foothills and the coastal Batinah region. Such plains are home to extensive *Acacia* woodlands (Fig. 3a) satisfying their water demands from the subterranean groundwater streams. These are the base of the sophisticated falaj water irrigation system which meets the needs of fresh-water requirements for the local population and their agriculture.

Falaj refers to a manmade water system that transports the underground water to the surface. The division of water from the falaj was very carefully regulated by the local farmers in the past, usually using an intricate system by the cycles of the moon. These systems can be seen all over Oman (also refer to EP 70).

This basal zone (Batanouny & Ismail 1985) is mainly dominated by an *Acacia tortilis-Euphorbia larica* woodland (Fig. 7, 8a) covering the altitudinal zone up to 1500 m. *Acacia tortilis* is spread all over Oman and one of the plants most favoured by camels, whereas the Rod Spurge *Euphorbia larica* is not affected at all. Frequent associates among the shrubs are Shaw's Teaplant (*Lycium shawii*) with tiny purple flowers and Sticky Fleabane (*Pulicaria glutinosa*) with yellow flower heads. The myrrh *Commiphora wightii* and the evergreen *Maerua crassifolia* of the caper family may be found in the lower-, the yellow flowered *Acridocarpos orientalis* (Malpighiaceae) in the upper range.

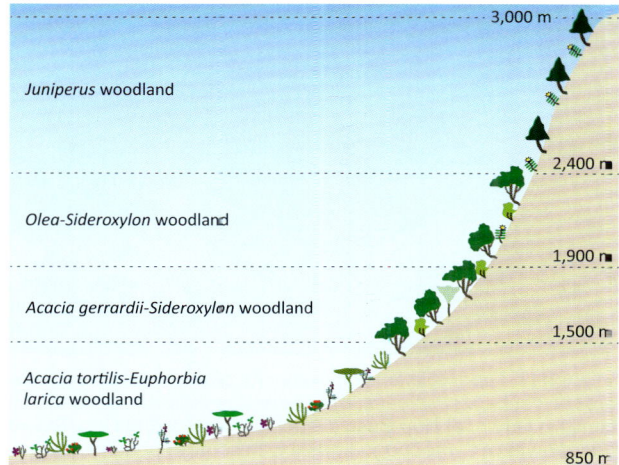

Fig. 7: Vegetation zonation in the Al Hajar Mountains from Wadi Ghul to Jebel Shams.

From the altitude of 1500 m and up, the *Acacia gerrardii-Sideroxylon* woodland is found. Gerrard's Thorn (*Acacia gerrardii*) is a highland tree, prefering the wadis in the belt between 1500–2000 m. A common associate with a wide range of site conditions is the ironwood *Sideroxylon mascatense* (Fig. 8c), an Indo-Malayan relict species with disjunct distribution in the Hindukusch, in south Arabia and in north-east Africa. Despite spiny lateral branch-

Fig. 8: a) *Acacia tortilis-Euphorbia larica* community of the northern foothills. b) African Olive (*Olea europaea* subsp. *cuspidata*). c) *Sideroxylon mascatense* is the only member of Sapodilla family in Oman. d) Hop Bush (*Dodonaea viscosa*). e) *Olea-Sideroxylon* zone with dense undergrowth of *Cymbopogon* grass with somewhat curled leaves. f) *Euryops arabicus* is the only species of a southern African centered genus radiating to southern Arabia.

es, it is commonly seen browsed down by goats to hedge-like shrublets looking like trimmed boxwood.

The altitudinal belt from 1900 to 2400 m is characterised by an evergreen *Olea-Sideroxylon* woodland (Fig. 8e) accompanied by the Hop Bush (*Dodonaea viscosa*; Fig. 8d) and the Buckthorn *Sageretia thea*. The African Olive (*Olea europaea* subsp. *cuspidata*; Fig. 8b) is an Afromontane element ranging to the Eritreo-Arabian region through eastern and southern Africa; the tree is often heavily browsed by goats. The lemon-scented *Cymbopogon* grass covers the understory.

Juniperus woodland (Fig. 3c) dominates from 2400 m upwards and forms an open woodland in the Hajar highlands. *Juniperus excelsa* subsp. *polycarpos* is an Irano-Turanian floral element and extends from Oman to east Turkey, central Asia and Pakistan. In the undergrowth, Rock Rose (*Helianthemum lippii*), Arabian Golden Daisy Bush (*Euryops arabicus*; Fig. 8f), Joint Fir (*Ephedra pachyclada*), Germander (*Teucrium mascatense*) or the Primrose *Dionysia mira* may be found.

Jebel Shams and the Saiq Plateau offer good observation opportunities of this altitudinal zonation described above and are also easily accessible (Ghazanfar 1991).

4.4 Dhofar Mountains

The Dhofar Mountains form a world of their own (Miller & Morris 1988). Influenced by monsoonal summer rains (Fig. 9), most of the landscape turns into a bright green when the *Khareef* arrives (Fig. 3f), and envelopes the wind-facing slopes in fresh drizzly weather from June to September. The monsoon forest at altitudes of (200–)300–900 m reawakens, forming a dense vegetation carpet characterised by the south Arabian endemic *Anogeissus dhofarica* (*Combretum* family; Fig. 10b) and dominated by an *Acacia-Commiphora* woodland. By the end of the year, most plants shed their leaves again to reduce evaporation during the dry winter season (Fig. 10a). Evergreen species like the ebony *Euclea racemosa* subsp. *schimperi* contribute to vegetation at climatically and edaphically most favoured sites. The succulent Bowstring Hemp *Sansevieria ehrenbergii* covers patches of rocky outcrops. The scattered *Acacia* woodland of the coastal plain is mostly degraded due to logging and overgrazing.

On the downwind side of the mountains, the precipitation and fog influence rapidly decreases, giving place to the perennial Rooigras *Themeda quadrivalvis* (Fig. 10c), characterising steppe-like grassland with some widely spaced trees of the Fig *Ficus vasta* (Patzelt 2011).

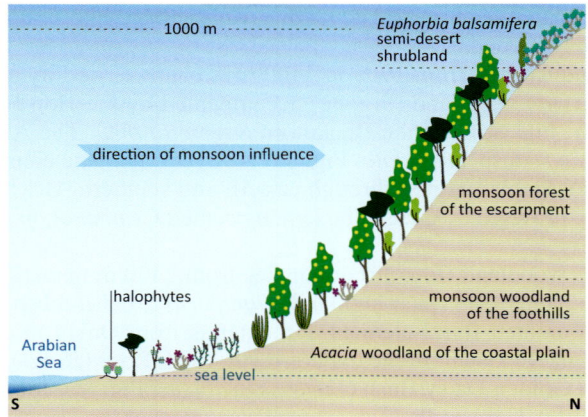

Fig. 9: Vegetation zonation at Jebel Qamar from sea level to 1000 m.

More desert-like areas in the rain shadow of the mountains are covered with the spurge *Euphorbia balsamifera* (Fig. 10d, e), forming hemispherical shrubs; the landscape resembling parts of the Canary Islands. The remarkable disjunct distribution area, indeed, covers Macaronesia and north-east Africa through south Arabia. Relatively dry regions are home of the Frankincense tree (*Boswellia sacra*); the locations formerly held top secret to keep the prices high and exclusive.

Livestock holders are plentiful, as are cattle and camels, and all of the landscape, except for the steepest slopes, is heavily browsed. Consequently, various unpalatable species take over. Thus, Bitter Apple (*Solanum incanum*), Apple of Sodom (*Calotropis procera*; Fig. 10f), the Physicnut *Jatropha dhofarica* and Veld Grape (*Cissus quadrangularis*) are among the common pasture weeds.

Among the invasive plants, the Mesquite tree (*Prosopis juliflora*; Fig. 10g) is worth mentioning. Formerly extensively introduced as a landscape plant, the species has recently turned into a serious pest common in Dhofar and in the

---

Fig. 10: a) Monsoon forest in the dry winter season. b) The Knob Tree *Anogeissus dhofarica* is a typical constituent of the deciduous monsoon woodland. c) *Themeda quadrivalvis* tall-grass savannah forms an orographic grassland in Dhofar. d) Dry season aspect of *Euphorbia balsamifera*, Jebel Samhan. e) *Euphorbia balsamifera* greening. f) Apple of Sodom (*Calotropis procera*) is an indicator of overgrazing. The poisonous latex makes the plant unpalatable. g) The Mesquite tree (*Prosopis juliflora*) is an invasive introduction from the Americas.

Batinah of northern Oman as well. The Mesquite is well adapted to drought and salinity, and mostly rejected by local livestock.

4.5 Land use

At higher altitudes with sufficient precipitation, terraced field cultivation is practised mostly around 1800–2000 m. Wheat and barley are among the main crops, Damask Rose (*Rosa* × *damascena*) is locally important for rose water production. Further on, Grape (*Vitis vinifera*), Walnut (*Juglans regia*), Apricot (*Prunus armeniaca*), Pomegranate (*Punica granatum*) are cultivated in a larger scale. In the medium course of mountain wadis, fields are encircled by small earth walls and irrigated by flood water through inundation.

Streams with permanent water are a rare phenomenon in a desert country like Oman. Wadi Tiwi or Wadi Bani Khalid are among the well-known sites with perennial surface water; the latter collected and transported by the famous falaj system described in Ch. 4.3 (Fig. 11). The shores are often fringed by Oleander shrubs (*Nerium oleander*; Fig. 12a) and appear in white to pink when in blossom. Among the swamp and water plants, Water Hyssops (*Bacopa monnieri*), Stiff Rush (*Juncus rigidus*) and tussocks of Plume Grass (*Saccharum*) are common.

In the desert, nature's resources are capitalised by grazing with camels, goats and sheep for meat and milk production. The Date Palm (*Phoenix dactylifera*; Fig. 12b) is the basis of economy in the oasis, where underground water

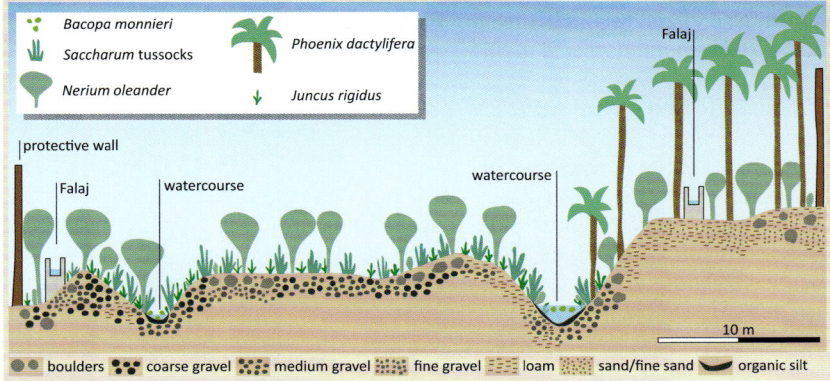

Fig. 11: Causal profile of the upper reaches of Wadi Bani Khalid in the eastern Al Hajar Mountains at about 650 m a.s.l.; topography not to scale.

Fig. 12: Rosebay (*Nerium oleander*). b) Date palms in the oasis of Al Kamil c) The peeling bark is a typical feature of *Boswellia sacra*, the Frankincense tree.

enables plenty of plant growth and fits the special requirements of that multi-usage tree which may be aptly described as "with head in sun and feet in water". The plants are dioecious, that means male and female flowers are produced on separate specimens. The farmers hold mainly females, which are pollinated by the wind, but for high quality varieties, male inflorescences are sold on the local markets and arranged into the females. At lower storeys, vegetables and other fruit crops are grown.

In the Dhofar coastal plain, coconut groves (*Cocos nucifera*) are an important source of income. The *Boswellic sacra* (Fig. 12c) populations in the hinterland are the basis for the frankincense resin trade and it is sold in quantities on the Salalah market and in markets all over Oman. For the production of frankincense, the outer bark is cut and the resin outflow collected after some weeks of drying. This aromatic resin has many health benefits; it can be chewed or dissolved in water. It also plays an essential role in Omani hospitality. When the guest arrives, a receptacle of burning frankincense is brought and the smoke is blown over the guest's clothes to refresh and revive the traveller.

5 Geology of Oman

5.1 Geomorphology

Oman is located at the eastern margin of the Arabian plate adjacent to active plate boundaries in the Indian Ocean and the Sea of Oman. With the rest of Arabia, Oman is currently being pushed slowly northwards at a rate of about 2 cm/year as the Red Sea grows wider. Through geological times, Oman's coastal margin has been deformed, uplifted, weathered, eroded and subsided under the sea a number of times making the environment in Oman so diverse and the landscape very colourful. Oman reveals most, if not all, of the main Earth-evolution chapters. From the highest peak of Jebel Shams in north Oman, 3 km above sea level, to the deepest point in the Sea of Oman, 3 km below sea level, a wide variation of topography and weathering profiles occur. These are a direct consequence of a rich geological history that produced different rock types and topographic profiles.

Oman has two prominent mountain chains. The northern one is known as the Al Hajar Mountains. Extending from Musandam to Ras al Hadd for about 700 km, this arcuate mountain chain is bordered by huge alluvial fans to the north and south. The Al Hajar Mountains contain the best exposed oceanic plate material in the world, known as the Samail Ophiolite. Its sequence contains different igneous layers of the oceanic crust and mantle that often weather into sharp and dark low-relief hills. The first layer from the top is of pillow lavas (see EP 89). These form when lava emerges on the sea floor and quickly cools down. There are vertical sheeted dykes below the lava units that resemble the original position of the fissure in the rock into which the magma surged (see EP 86). The deeper units of grey layered gabbro, a dark, coarse and grainy rock, and brown mantle rocks form the deeper part of the sequence. These are separated by the Moho (short for Mohorovičić discontinuity) that divides the crust from the mantle and normally lies 8 km below the sea floor (see EPs 57–60). The ophiolite was emplaced on the passive margin of Oman along with deep ocean sediments, known as the Hawasina rocks. They form thinly-bedded chert units that can be seen along many road cuts in north Oman (see EPs 96–99). Large pale "exotic" rocks typical of shallow water were formed above volcanic islands in the ocean (see EP 35). As they were transported from the ocean to Oman, the rocks were highly deformed, faulted and folded.

The Batinah coastal plain in the northeast of the country is composed of gravel and sand, eroded from the adjacent mountains. A shallow aquifer allows for agricultural use on the outskirts of the plain. The debris of the oceanic rocks cover the coastal plain of Al Batinah. The plain is narrow but extends for about 200 km. Many wadis from the Al Hajar Mountains cut across the plain. There are also a number of linear sand dunes formed by the northern *Shamal* winds. The oases are supplied by the underground water. 3 % of Oman's land can be considered as coastal plain.

The high-relief massifs of the Al Hajar Mountains are mainly made up of limestone units. In the Musandam peninsula that lies on the northern tip of Oman, the limestone is predominantly of Mesozoic age. Close to Musandam, the collision between Arabia and Eurasia occurs. Because of this, the mountains in Musandam subside by around 7 mm every year. Rocks buckle and break, mountains form and get quickly submerged under the sea. Today, ancient valleys and caves are sea fjords where a diverse range of submarine life flourishes.

The Jebel Akhdar massif in the central part of Al Hajar Mountains separates Al Batinah from Al Dhakhiliyah governorates. Glacial rocks were deposited here 700 million years ago. The view from Jebel Akhdar towards the gravel plains in Al Batinah to the north and Al Dhakhiliyah to the south is breathtaking. On the eastern side of the Al Hajar Mountains lies Saih Hatat. It is about 70 km by 50 km wide. The mountain is deeply eroded and has a moon-like landscape predominantly made up of schist, which is a medium grained metamorphic rock. On its flank, the contact between the ophiolite from the former oceanic lithosphere and the limestone of the continental crust is clear.

Along the contact line between the Samail Ophiolite and limestone of Jebel Akhdar and Saih Hatat, lie a number of hot springs, such as Al Kasfah in the Wilayat of Ar Rustaq or Ayn A' Thowarah in Nakhl (see EP 51). These springs emerge from considerable depths along a fault, known as the frontal range fault that separates the two rock masses (see EP 50). The temperature of these springs varies between 40 to 70 degrees C. The hottest of these is Ayn Al Hamman in Muscat (UTM 40 Q 635204 2596193/ N 23°28'12" E 58°19'25").

The limestones of Jebel Akhdar and Saih Hatat have a number of large caves that have primarily formed because of limestone dissolution (karstification). In fact, the Tertiary limestones that crop out on the eastern end of the Al Hajar Mountains include a number of the largest caves in the world. Kashlat Meqandili or Majlis Al Jin, is the largest cave in the area and among the top 10 in the world (see EP 12). It is a single chamber measuring 310 meters by 225 meters with a dome ceiling 120 meters high and three entrances on the roof.

The wadis on the eastern part of the Al Hajar Mountains combine coastal beauty with inland splendour. From Wadi Tiwi (EP 18), it is a day's hike to the other side of the mountain, to Wadi Bani Khalid, well-known for its springs, water pools and underground caves. If you are interested in hiking, we recommend the book by Dale & Hadwin (2001) where, the track from Wadi Tiwi to Wadi Bani Khalid is described (p. 222ff).

A second mountain chain is located in the Dhofar province in the south of Oman, bordering the Republic of Yemen. This chain is composed of three main mountains. The northern one is Jebel Samhan, where the Arabian Leopard Sanctuary is situated. Wind and water erosion cast the Tertiary flat beds of the crest. In the foreground, crystalline rocks crop out on the low coastal plain. The lines running along the plain are igneous dykes cutting across older rocks and were formed as magma was forced up through fractures. Dhofar Mountains are well watered. Thick travertine limestone beds are formed in many wadis. During the summer, the south-eastern monsoon covers the area with lush green vegetation, particularly in Jebel Qara. Toward Jebel Al Qamar, massive cliffs created by faults form the sharp southern boundary of the slope. To the north, the Dhofar Mountains have a gentle slope. A number of wadis, such as Wadi Dawkah, flow to north.

Masirah, the largest island in Oman as well as Arabia, lies on the eastern side of Oman, around 24 km from the coast. The island is mainly covered with ophiolite that was generated around 140 million years ago in what is known today as the Arabian Sea. The Masirah Ophiolite was emplaced here around 65 million years ago, after the Samail Ophiolite in north Oman. The Huqf area is located west of Masirah on the main onshore side of Oman. This area provides a window onto a geological history spanning more than 700 million years. It is another of Oman's unique natural museums. As the area is remote, only a few stops along the main road are included in this book (see EP 45, 46, 48).

Desert plains form most of Oman's landscape. These are covered by wadi gravels in north and south Oman. Alluvial fans emanate from the Oman Mountains and reach central Oman. The longest is Wadi Andam which extends for 209 km. Wadi Andam erodes older wadi fans that have flowed during wetter climate periods in the past. Northern Oman wadis that flow southwards terminate along an area of subsidence in central Oman which appears white from space because of evaporates that precipitate in here. This place is known as Umm as Samim Sabkha and forms the western side of this zone, whereas Bar Al Hikman is located on its eastern side. Umm as Samim, translated as Mother of all Poison, is the largest *sabkha* or *playa* in Oman. As wadis flow from mountains, they pick up salt from rocks. As the water evaporates, the salt forms a polygonal-ridges crust several meters thick. The Bar Al Hikman peninsula is a large salt pan with tidal lagoons that attract different types of birds. It works as a perfect laboratory to understand carbonates and evaporates deposition. The desert plain in central Oman and along the south-eastern coast has a limestone plain (see EP 48) covered with rubbly sediments and some patchy dunes (see EP 47).

Oman has two main sand dune bodies. From space, the Wahiba and Sharqiyah Sands give the appearance of being an arm muscle (see EP 32). The northern edge of the Sharqiyah Sand is truncated by Wadi Batha. Isolated masses of

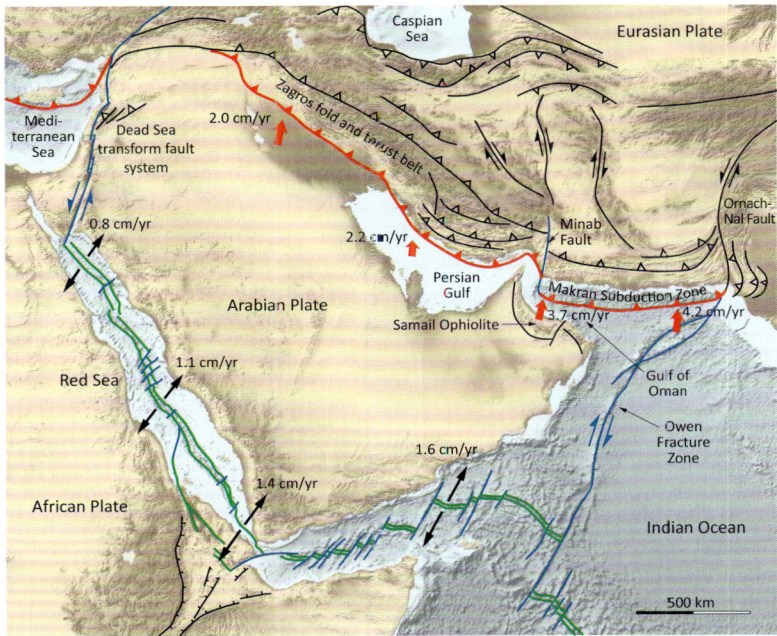

Fig. 13: Plate tectonic setting of the Arabian Plate. Plate motion velocities after Kukowski et al. (2000).

dunes are present on the other side of the wadi. A single linear dune of the Sharqiyah Sand can be more than 100 km long and 100 m high. Most of the dunes were formed by the former strong south-western monsoon that used to deflate exposed sea floor in the southern side of the Sharqiyah Sand during ice ages that occurred earlier than 14,000 years ago. To the south, the sand dunes become smaller and at the coastline, they are cemented hard, probably because of high content of carbonate grains (see EP 43). The Rub al Khali, Empty Quarter, is the main sand body in Arabia. It forms the south-western part of Oman. At the edge of the Rub al Khali, the sand dunes are large and separated by *sabkha* because of the sand and water supplies from wind and water. The dunes of the Rub al Khali to the north are mainly star-shaped (see EP 47) because winds rapidly change direction here.

Both the eastern and northern continental shelves of Oman are narrow and steep. They descend 3000 m to the ocean floor. Sediment slumps frequently fall off this slope. The widest shelf which extends for almost 100 km is seen in the central part of the eastern coast of Oman. The northern and eastern parts of Oman are bound by crustal fracture zones. The eastern side of Oman is bordered by the Owen Fracture Zone, which represents the boundary of Arabia and the Indian subcontinent (Fig. 13). To the south, this zone connects with the Gulf of Aden and the Red Sea spreading ridges. In contrast, the Makran and Zagros collision zones represent the northern border of Arabia. The Makran Subduction Zone is located at the northern end of the Gulf of Oman. Here, the sea floor and oceanic crust of the Gulf of Oman is very active and gets consumed below the continental crust of Makran, possibly at a rate of about 4 cm per year. Today, the narrow oceanic crust of the Gulf of Oman represents a remnant of a massive old ocean, known as the Neotethys.

5.2 Major tectono-stratigraphic units

In general the geological record can be classified into two main groups: 1. rocks that formed in the place where they can be studied today, and 2. rocks that were transported as thrust sheets or nappes due to the application of tectonic forces in compressional settings. The first group is also referred to as autochthonous units, where the term *autochthon* derives from ancient Greek and can be translated as *indigenous*. In contrast, the term *allochthon* literally translates to *other ground*, meaning *foreign* in a wider sense. Hence, these rocks are referred to as allochthonous units.

The entire geological history of Oman may be classified by the application of this system, comprising seven major rock units:

1. The crystalline basement

This rock unit formed by continental accretion during the Late Proterozoic. This means a process of growth, facilitated by the gradual accumulation of additional layers of material. It comprises of metamorphic rocks like gneiss and micaschist that are intruded by various types of igneous rocks like dolerites, granodiorites and granites.

There are limited surface outcrops in Oman, one example being at Jalan Bani Buhassan (EP 31). The basement is also penetrated by 5 wells (Loosveld et al. 1996). Radiometric age dating reveals the formation and cooling of the rocks in the period 830 to 730 million years ago (Gass et al. 1990). The same cratonic rocks are known in outcrops from the Arabian Shield further west, indicating that the Arabian plate is inclined towards the east. The term 'craton' refers to the more stable parts of the continental crust, differentiating them from a more geologically active area.

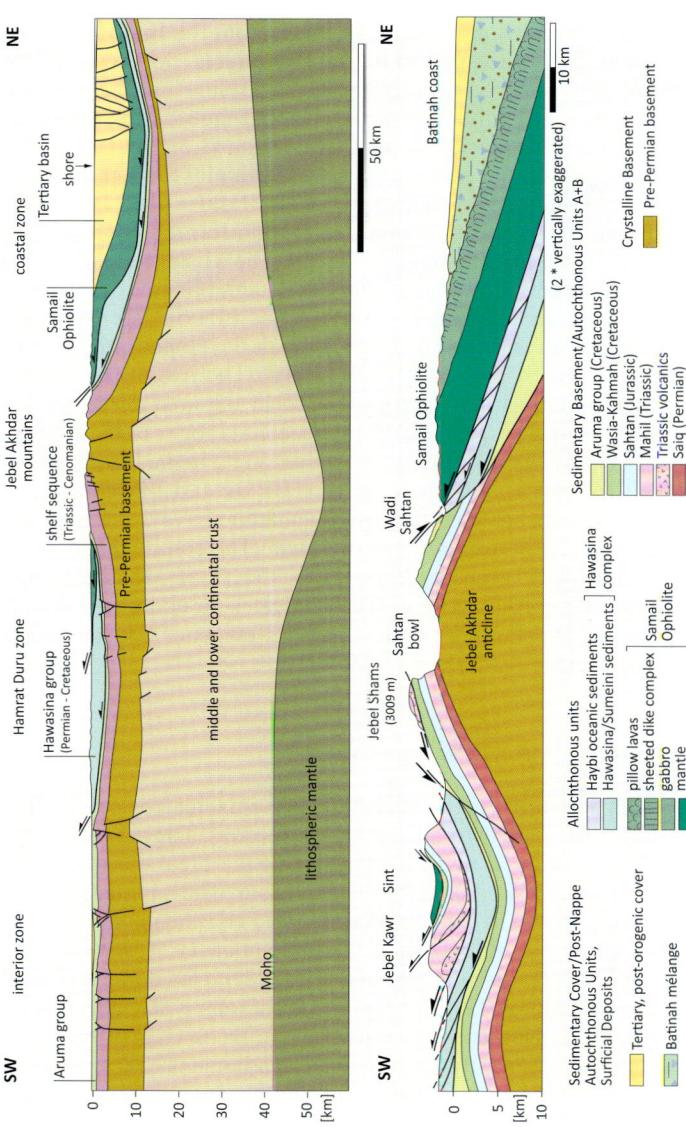

Fig. 14: Schematic cross section through the main structural units of northern Oman (above) and in detail the Jebel Akhdar dome (below; modified after Al-Lazki et al. 2002).

2. Sedimentary Basement/Autochthonous Unit A

The rocks integrated in this unit were laid down from the End Proterozoic to the Early Permian. Oman was part of the supercontinent Gondwana and sedimentation is mainly restricted to intercontinental basins and the shelf of the continent during this long time period. Extensional fault basins developed in the Precambrian/Early Cambrian in southern Oman (Loosveld et al. 1996). These northeast-southwest trending basins were initially predominantly clastic, referring to rocks composed of broken pieces of older rocks (Abu Mahara formation) passing into platform carbonates (Nafun formation). They unconformably overlie the basement; compare Fig. 14. The Mistal formation (see EP 55) in the Jebel Akhdar is stratigraphically equivalent (Glennie et al. 1974) as is the Hatat formation in Saih Hatat (see EP 05 and EP 06). The asymmetrical South Oman Salt Basin and the Ghaba Salt Basin developed, filled with Ara group evaporates (Terken et al. 2001; see EP 49). The rest of the Palaeozoic until the Permian is only fragmentary exposed in the area described in this fieldguide (cf. Millson et al. 1996). Exceptions are the rocks of the Lower Palaeozoic (mainly Ordovician) Haima group outcropping in the Huqf (see EP 45) and the Amdeh formation in Saih Hatat (see EP 07 and EP 09). Terrestrial conditions on the Gondwana continent are assumed to have been predominant during the Devonian and Carboniferous of Oman. The continental deposits of the Permian Gharif formation are exposed in the Huqf Mountains but are pinching out towards northern Oman. This indicates an updoming during the Late Carboniferous and Permian and is seen as evidence for the following breakup of Gondwana (Blendiger et al. 1990).

3. Arabian Platform/Autochthonous Unit B

Most of the rocks that form the mountains in northern Oman belong to this unit that formed during the Late Permian to Late Cretaceous. The passive margin of the Arabian Plate developed as a consequence of the break-up of Gondwana and the establishment of the Neotethys ocean basin. The submergence of the Arabian Platform initiated in the Mid-Permian and continental debris is transgressively onlapped by marine deposits (Glennie et al. 1974; see EP 61). The Mid Permian to Early Triassic carbonate platform deposits make up the Khuff formation (Koehrer et al. 2012). The extensive carbonate platform that established itself upon this Neotethys continental margin is characterised by the sedimentation of predominantly shallow-water carbonates like limestone and dolostone, throughout most of the Mesozoic. Occasional pulses of clastic sediments are interpreted as eroded material from the exposed Arabian Shield (Droste & van Steenwinkel 2004).

The carbonate platform sedimentation (see EP 09) was finally terminated in the Cretaceous due to uplift. This vertical movement is seen as a consequence of the formation of a peripheral bulge. Eoalpine compressive tectonism resulted in subduction and later obduction, due to the collision with Eurasia (Warburton et al. 1990). The Late Cretaceous Alpine Orogeny is further responsible for the faulting and folding of the carbonate platform sediments, resulting in a westward dipping anticline represented by the Jebel Akhdar structure (Al-Lazki et al. 2002, Searle 2007).

4. Hawasina nappes/Allochthonous Unit

This unit is named after Wadi Hawasina, located in the north of the Al Hajar Mountains, where the associated rocks are exposed. The initial Permian rifting of Gondwana was soon followed by seafloor spreading and the establishment of an oceanic basin. Deep sea and continental slope deposits, mainly pelagic, from the open sea, and turbidite sediments, formed at the same time as the accumulation of shallow water deposits on the passive margin of the Arabian Plate during Late Permian to Late Cretaceous (Arabian Platform/Autochthonous Unit B). This ocean basin is called the Hawasina basin and forms part of the southern margin of the Neotethys Ocean. Oceanic basin sediments are also summarised as Haybi. The basin has a characteristic seafloor topography including continental slope, seamounts and mid-oceanic ridges. Different rock types formed here and are grouped accordingly.

The proximal (near to source area) Sumeini group sediments accumulated along the continental slope (cf. Fig. 14). In the deeper parts of the basin, distal deep ocean sediments were laid down. Typically these basins are so deep that $CaCO_3$ dissolves and only fine grained terrestrial material as well as the remains from organisms made up of SiO_2 (e.g. radiolarians) accumulate. The resulting rocks are radiolarian chert (see EP 24).

The deeper part of the ocean basin is further subdivided into a more proximal basin, called the Hamrat Duru basin and a more distal section, called the Umar basin (see Hanna 1995, and compare Fig. 14). The rocks of the Kawr group intervene and separate the basins. This group comprises of limestones as well as volcanic rocks that make up the Oman exotics (see EP 35 and EP 81). Typically they are composed of shallow marine reef limestone. The volcanic substratum may be interpreted as a guyot, an isolated submarine volcano, indicating an atoll-like setting. They are Mid-Permian to Triassic in age. Often, the rocks are slightly metamorphosed, resulting from the overriding by the ophiolite. The noticeable quarry activities on these light yellowish mountains are a consequence of marble mining.

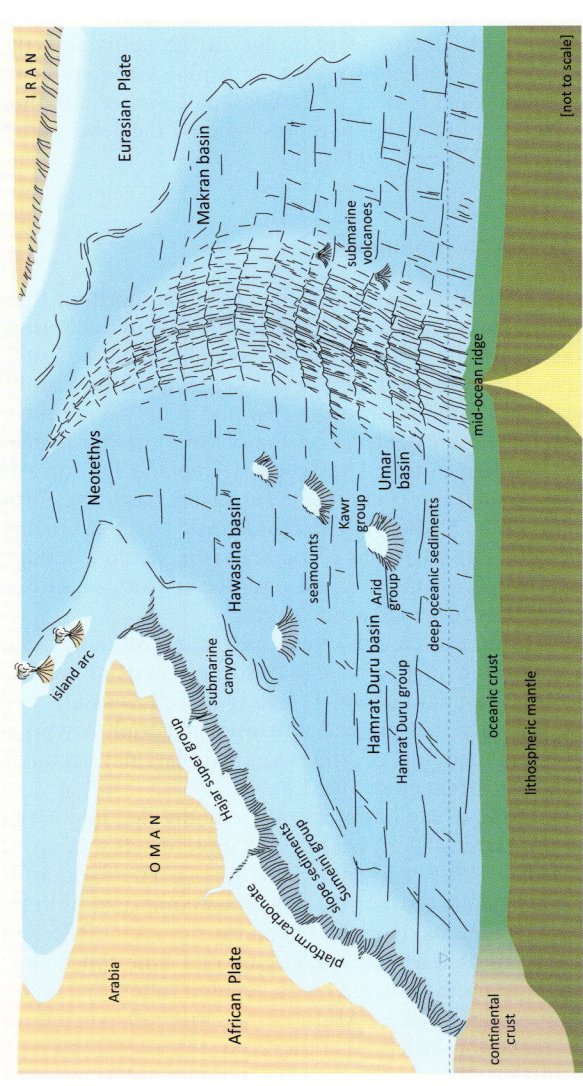

Fig. 15: Seafloor topography of the Neotethys Ocean in the Cretaceous. Autochthonous units formed on the continental crust, allochthonous units are represented by sediments on oceanic crust and the ophiolitic rocks of the Samail Ophiolite. Modified after Hanna (1995).

This entire rock sequence is allochthonous and the rocks were pushed onto the edge of the Arabian plate. This compressive tectonism resulted in extensive rock deformation which can be studied in the outcrops (e.g., EP 24, EP 25, EP 80, EP 97–EP 99).

5. Samail Ophiolite/Allochthonous Unit

These igneous rocks formed during the Middle to Late Cretaceous. The spreading of the seafloor in the center of the Neotethys Ocean basin resulted in the formation of oceanic lithosphere and the widening of the basin, later followed by an intra-oceanic subduction. This oceanic lithosphere is referred to as ophiolite when exposed on land. Today the rocks from the Samail nappe form the structurally highest allochthonous units. The Samail Ophiolite represents the best exposed oceanic lithosphere worldwide. See Excursus I for a detailed description and the process of formation.

6. Sedimentary Cover/Post-Nappe Autochthonous Unit

Northeastern Oman was subject to marine conditions again during most of the period from the Late Cretaceous to the Neogene following the emplacement of the nappes. However, Jebel Akhdar and Saih Hatat must have been sub-aerially exposed as evidenced by onlapping Upper Maastrichtian and Paleogene sediments (Searle 2007). Locally, terrestrial sediments accumulated. Post-orogenic clastic delta sediments dating to the Late Cretaceous are recorded from the Al Khoud area close to Muscat. Vertebrate fossils including dinosaur bones from these conglomerates are described by Schulp et al. (2000, 2009).

A marine transgression with the establishment of shallow-marine sedimentation resumed at the very end of the Late Cretaceous and persisted until the Oligocene, interrupted by emergence caused by global sea level fall (Searle 2007). The dominating rocks are marl deposits which are exposed e.g. along the Muscat Express Way (see EP 40). Northern Oman underwent a complex history of Cenozoic tectonics (Fournier et al. 2006) before compression finally resulted in the uplift of the Oman Mountains. Cenozoic refers to the last period in geological history, from around 65 million years ago up to the present.

7. Surficial Deposits/Post-Nappe Autochthonous Unit

During the Quaternary the landscape of modern Oman was shaped by the changing influence and interplay of climate and tectonics in combination with variations in global sea level. The tectonically triggered uplift of the mountains led to the erosion of the sub-aerially exposed rocks. The climate of the Arabian Peninsula changed as the world's climate was in the process of shifting from a cold to a warmer phase. The immediate impact was related to shifts of the monsoonal belt and associated wind systems. Cold phases were responsible for a drop in sea-level and exposure of the continental shelf.

Wind was able to entrain or transport sediments, e.g. the sand that makes up most of the dunes in the sand seas (see EP 32 and EP 43). Wet phases were responsible for intensive weathering, especially as the dominating rock type is limestone, susceptible to dissolution. This process is also known as karstification and led to the formation of large cave systems (see EP 12). Activation of river channels during wet phases led to the transport of vast amounts of weathered material into the foreland of the mountains. Huge alluvial fans formed and merged, and today make up the Batinah coast in the north. The ophiolite rocks, on the other hand, reacted to sub-aerial conditions by rock alteration (see EP 60). The landscape dynamics, in response to changes in climate, tectonics and sea-level, can still be traced along the slopes of the mountains (see EP 11).

5.3 Plate tectonic evolution

The Arabian Plate is bound by the strike-slip fault zone of the Owen Fracture Zone to the south-east representing the boundary between the Arabian and Indo-Australian Plate. This zone is characterised by active, dextral (on the right hand side) strike-slip motion (Fournier et al. 2008, 2011). The Arabian Plate is further confined by the active spreading axes of the Gulf of Aden and the Red Sea to the south and west, separating it from the African Plate, and by the Dead Sea transform fault to the northwest (Hempton 1987, McClusky et al. 2003). The northeastern boundary is marked by a continent-continent, partly continent-ocean collision zone which results in the Zagros and Makran fold and thrust belts (Ross et al. 1986, Vernant et al. 2004).

Fig. 16: Plate tectonic evolution of the northwestern branch of the Neotethys Ocean from the Triassic to Middle Tertiary.

Plate tectonic evolution

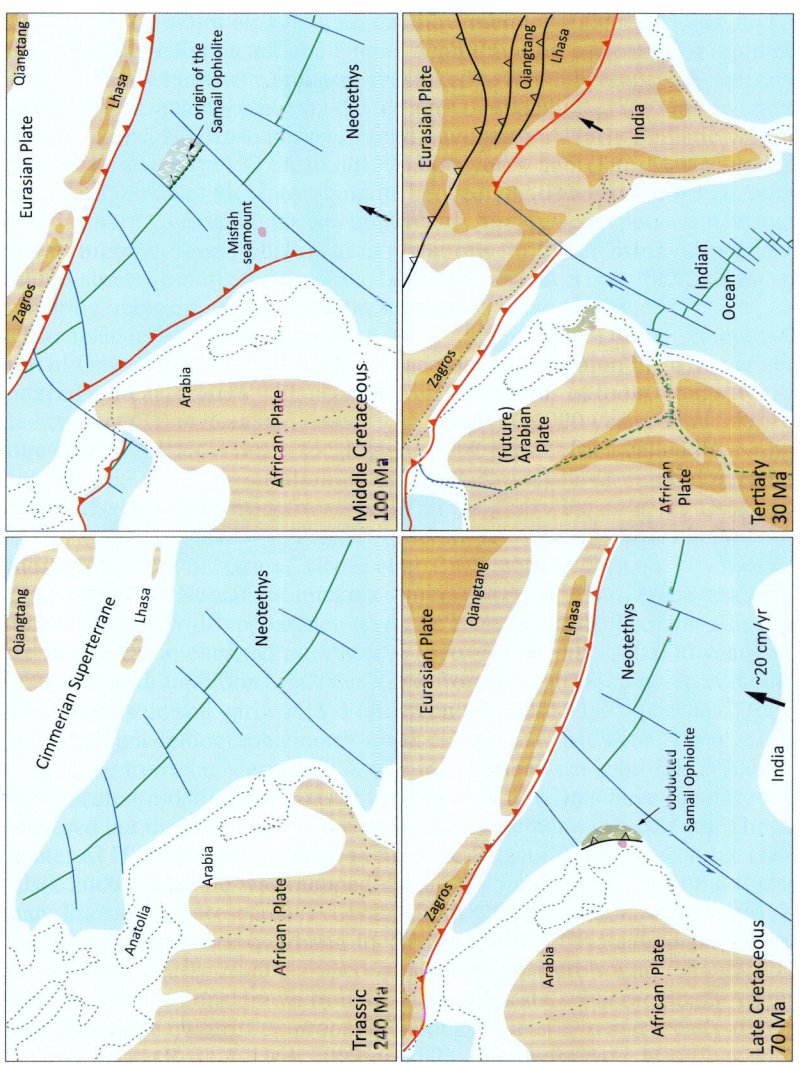

The Makran Subduction Zone (MSZ) is situated within the Gulf of Oman trending east-west and representing a subduction zone which has been active since the Miocene (Glennie et al. 1990) or possibly since the Early Cretaceous (Byrne et al. 1992). It stretches from the Minab Fault System in the Strait of Hormuz, to the Ornach-Nal Fault in Pakistan and has an along-strike extension of about 1,000 km (Page et al. 1979, Smith et al. 2012). White (1977) suggested that the Arabian Plate dips by an angle of 5° towards the north (010) beneath the accretionary prism or wedge. The convergence velocity is estimated by Regard et al. (2005) and Copley et al. (2010) to an average of 40 mm/year in the east (see Fig. 13). The MSZ has one of the largest accretionary wedges observed at convergent margins with a very high sedimentary input (Kopp et al. 2000, Prins & Postma 2000, Bourget et al. 2013). The oceanic crust which is currently being subducted in the MSZ is a remnant of the Neotethys Ocean. Based on differences in seismicity, the MSZ is subdivided into a western and an eastern part, where the eastern part is characterised by large thrust earthquakes, while the western part has experienced no great earthquakes in historic times (Byrne et al. 1992, Smith et al. 2013). This plate tectonic setting is also indicated by the complex tectonic geomorphology in the Sea of Oman (Uchupi et al. 2002).

Over a long period of time, the Arabian Plate belonged to the African Plate. It was separated from the African Plate when the Dead Sea rift system started to open in the Oligocene (Lyakhovsky et al. 1994). The final separation from the African Plate occurred, however, not before the Early Pliocene, about 5 million years ago. At this time the Dead Sea transform fault came into existence as a consequence of the change in movement of the Arabian Plate. During Permo-Mesozoic times the area of Oman was located at the western passive margin of the Neotethys Ocean.

The situation changed in the Middle Cretaceous when the Arabian promontory of the African Plate collided at its northeastern boundary with the oceanic crust of the Eurasian Plate. This collision occurred along the mid-oceanic spreading center of the Neotethys (Fig. 16) and initiated the formation of the ophiolite nappe, which became the Samail Ophiolite, when it was obducted onto the continental margin of the Arabian part of the African Plate (Fig. 17). The obduction was completed by the end of the Cretaceous. The collision went along with the very fast lateral movement along the Owen Fracture Zone and the convergence between the African and Eurasian plates. During the Cretaceous and Early Tertiary, the Indian continent moved with a velocity of about 18–20 cm/year in a northerly direction, finally colliding with Eurasia in the Late Paleocene/Early Eocene (Kious & Tilling 1996). The convergence rate was then reduced to the present 4 cm/year (Copley et al. 2010).

The tectonic unit below the ophiolite nappe, the former continental margin of the African Plate, is today represented in the Saih Hatat dome. Folded nap-

Plate tectonic evolution 43

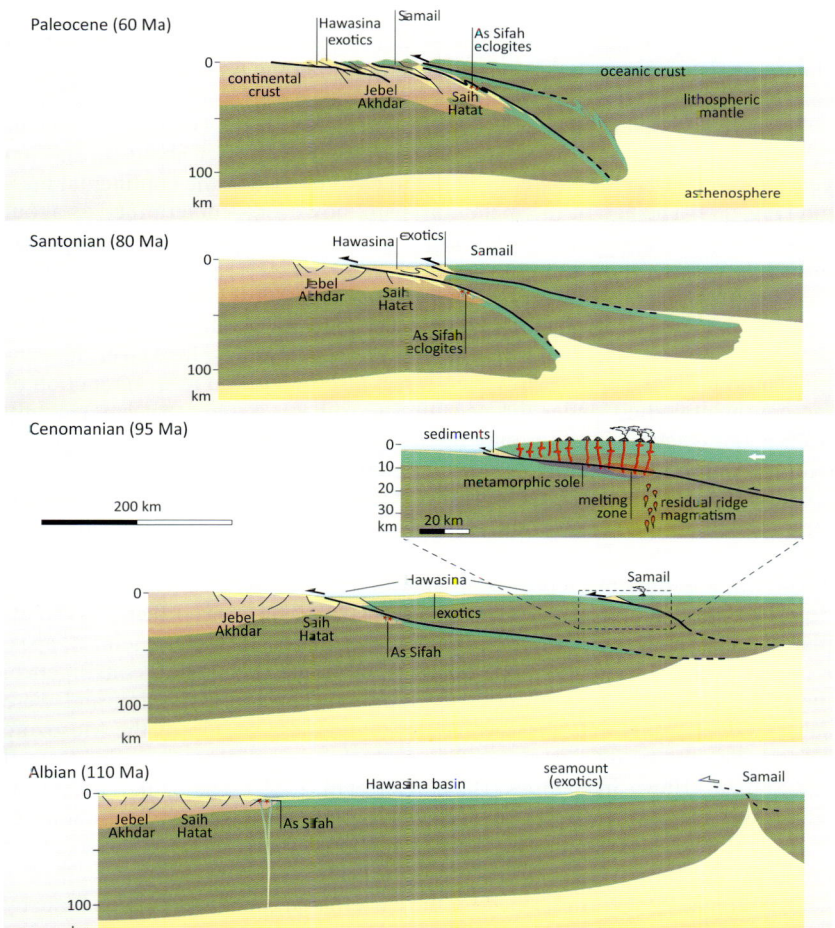

Fig. 17: Obduction of the Samail Ophiolite and formation of the Samail nappe as a consequence of convergent plate movement between Eurasia and Arabia (modified after Nicolas 1995, Frisch et al. 2011, Frisch & Meschede 2013).

pes of Pre-Permian and Mesozoic metamorphic rocks occur below Cretaceous blueschist and eclogite metamorphic rocks which are overthrust onto the Saih Hatat unit. Blueschists and eclogites are exposed in the As Sifah region, southeast of Muscat on the coast of the Gulf of Oman. The age of the eclogites is radiometrically determined as 110 million years and is thus older than the emplacement of the ophiolite nappe which started to obduct about 95 million years ago. The eclogites are considered to have formed at the continental margin (Fig. 17). The age of the metamorphism, however, is determined as about 80 million years and is interpreted to have occurred at a late stage of the overthrusting process of the Samail Ophiolite.

The most prominent event in the geological evolution of Eastern Arabia was the emplacement of the oceanic crust onto the Arabian Platform. This process is mostly described as obduction of oceanic lithosphere over the continental margin as nappes. However, it can also be viewed as a subduction of the continental shelf under the oceanic lithosphere, a view that has recently been under discussion and it may be helpful to understand the later structural evolution of the Al Hajar Mountains (Glennie 2005).

However, the oceanic crust of the Neotethys began to override the continental margin at around 80 million years ago, after some older oceanic crust was consumed (e.g. Searle 2007). The overthrusting led to the imbrication of the deep-marine sediments deposited in the former ocean with volcanics and mantle rocks. First the sequences of the Hawasina basin were affected and emplaced as the Hawasina-Haybi complex, together with the Sumeini nappes, represented by continental slope deposits. With the proceeding convergence, magmatic crust and mantle rocks were affected by the obduction and were subsequently thrust over the continental margin and on top of the Mesozoic units during the Late Cretaceous. These allochthonous units are called Samail and Sumeini/Hawasina nappes (Béchennec et al. 1990). The Samail nappe encompasses the world's best exposed coherent ophiolite sequence (Boudier & Nicolas 1995). These ophiolitic rocks characterise the very rough landscape of medium elevation.

A similar process is described for the eastern margin of the Arabian Plate by Immenhauser et al. (2000). Here, the Batain basin formed during the Carboniferous/Early Permian period. The eastern part of the Arabian Plate, including the adjacent oceanic lithosphere of the Batain basin drifted northward during the middle Cretaceous. The oblique collision with the Indian Plate caused overthrusting of Lower Permian to Upper Cretaceous sediments, volcanic rocks as well as fragments of the oceanic lithosphere onto the northeastern margin of Oman. The latter are referred to as "eastern ophiolites" and the emplacement occurred approximately 15 million years after the Samail Ophiolite.

Post-nappe autochthonous units of late Cretaceous to Tertiary age that represent terrestrial and shallow marine environments cover the Samail Ophiolite. The huge gravel and sand deserts that dominate the interior of the country south of the Al Hajar Mountains, developed during the Quaternary. The main uplift of the mountains occurred during the Neogene as indicated by studies conducted within the bajada that borders the Al Hajar Mountains to the south (Blechschmidt et al. 2009). Bajada usually refers to a broad sloping area of alluvial material at the bottom of a mountain.

Because of the presence of obducted and, therefore, allochthonous rocks, the major geological units in northern Oman are divided into autochthonous and allochthonous units. The autochthonous rocks formed on the Arabian (African) continent or shelf and comprise successions before and after the obduction. The allochthonous rocks originally formed on the oceanic crust of the Neotethys Ocean (Fig. 15). They were dislocated by thrusting several hundred kilometers towards the south-west during the closure of the Neotethys and the obduction of the oceanic crust and lithosphere (Searle & Cox 1999). The overthrusts used the weak zones of spreading centers in the mid-ocean ridge (Glennie 2005).

Excursus I: Ophiolites

Definition

The name ophiolite comes from the Greek words "ophis" (ὄφις) and "lithos" (λίθος) which mean snake and stone, respectively. The term was first introduced by the French natural scientist Alexandre Brongniard in the 19th century for an assemblage of green rocks in the Alps (Brongniard 1813). In 1905 the German geologist Gustav Steinmann modified its use to a sequence with serpentinite, pillow lava, and chert, the so-called "Steinmann's trinity". In particular, the scaly surface of a serpentinite (Latin "serpens" = serpent, snake) resembles the skin of a snake. Although the rock association of ophiolites has been well known for a long time, the concept of ophiolites became newly relevant when the theory of plate tectonics was generally accepted by the scientific community in the 1970s. By then it became accepted that ophiolites represent parts of the oceanic lithosphere that had been emplaced on land. They are thrust onto the continental crust during orogenetic processes and at several localities, Oman being one of them; it is possible to study ocean floor material on land. These rare remnants are, therefore, critical evidence in the reconstruction of plate tectonic processes of the geological past.

Ophiolites are defined as a series of rocks that represent a cross section through the oceanic lithosphere. Starting at the base, an ophiolite is composed of peridotites which were originally formed in the uppermost mantle and transformed into serpentinites to different degrees by water absorption. In the uppermost layers ultramafic cumulates may occur. The next layer up will usually be gabbros which may partly be metamorphosed to amphibolites. Frequently, distinctly layered gabbros occur together with cumulates at their base. In some places, pods of plagiogranite occur at the top of the gabbroic layer in the gabbro-dike transition zone. Plagiogranite is a differentiated rock richer in silica than gabbro that was formed by the remelting of crust altered by hydrothermal fluids (Grimes et al. 2013). Above the gabbros, the next layer is the so-called sheeted dike complex, where vertical dolerite dikes intrude one another. It is overlain by basaltic layers, frequently with pillow or tubular basalts, and in places horizontal sills. In the interstices of the pillows, hyaloclastites are accumulated, which are composed of broken volcanic glass. The uppermost lay-

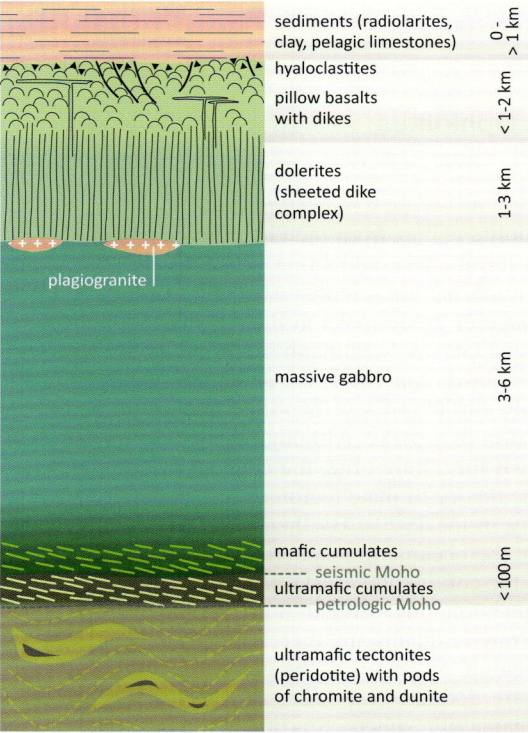

Fig. 18: Schematic columnar section of a typical ophiolitic sequence (modified after Frisch et al. 2011, Frisch & Meschede 2013).

er is represented by deep water sediments either carbonate-free radiolarites and clays or pelagic limestones (Fig. 18).

Normally, oceanic lithosphere is more or less completely subducted and recycled into the mantle at subduction zones. In some cases, however, parts of the ocean floor can be thrust as a tectonic nappe onto a continental margin by compressional tectonics. This process is called "obduction" (Latin "obducere" = to cover or to pull over). This is in contrast to subduction (Latin "subducere" = to lead away downwards) where the oceanic lithosphere is led downwards, deep into the mantle. Obduction particularly occurs when young ocean floor which is relatively hot and, therefore, less dense than normal ocean floor, is involved in the compressional processes because it is not yet able to subduct. Other ophiolites are trapped in collisional orogens when island arcs or continents collide with other continental blocks. In these cases, the suture zone between the colliding blocks is marked by commonly dispersed and deformed ophiolite remnants and the zone is called an "ophiolitic suture". This type of ophiolite is frequently characterised by high-pressure metamorphism (Ernst 1981) where basaltic and gabbroic rocks are metamorphosed to blueschists or eclogites, and ophiolitic melanges.

Ophiolites generally represent only a very small portion of the original ocean floor, since under normal conditions it is completely subducted into the mantle. Also, these remnants are strongly affected by the tectonic processes of subduction and/or obduction which result in often highly deformed and metamorphosed ophiolitic rocks. In most cases the original sequence of ophiolitic rocks is not preserved. In the Samail Ophiolite, however, the obducted ophiolitic unit was very large and thus remained intact in large portions. It is, therefore, one of the best exposed examples of an ophiolite on Earth (Glennie et al. 1974).

Formation of oceanic lithosphere

Following the text books of Frisch et al. (2011) and Frisch & Meschede (2013), normal oceanic lithosphere is approximately 80 to 90 km thick with 5 to 8 km of oceanic crust when it attains an age of approximately 80 million years. At the spreading center of a mid-oceanic ridge, however, the asthenosphere, which is the highly viscous and weak layer of upper mantle just below the lithosphere, rises nearly up to the surface directly beneath the newly forming oceanic crust. Here, the material of the asthenosphere has its highest degree of partial melting with 20–25 % melt in a fast spreading zone (Fig. 19). In a slow spreading zone the amount of partial melting is lower. Beneath the spreading center, the peridotite is split into a molten part of basaltic composition and a solid peridotitic residual rock. The basaltic melt rises up into a magma cham-

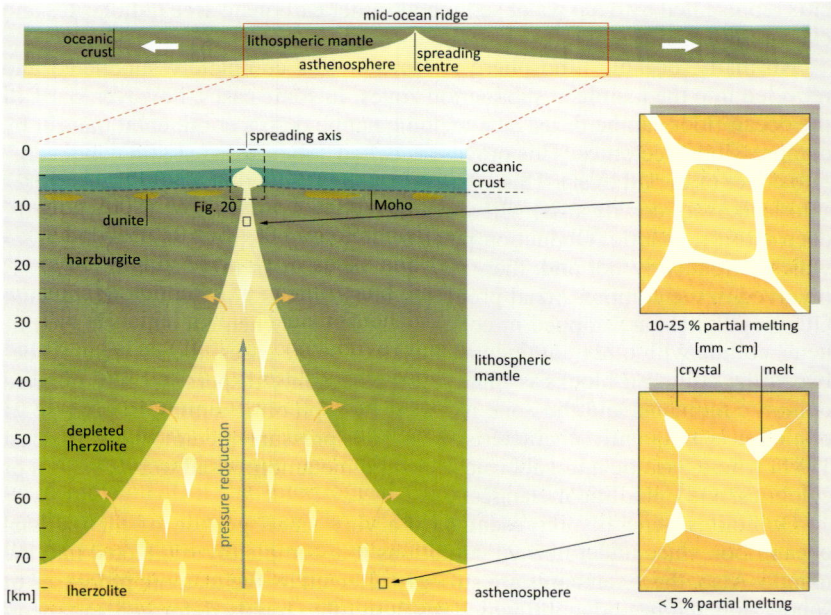

Fig. 19: Formation of oceanic lithosphere and melting processes at a mid-oceanic spreading center (modified after Frisch et al. 2011, Frisch & Meschede 2013).

ber which feeds the volcanoes at the spreading ridge, the sheeted dike complex, and solidifies at its flanks to gabbroic rocks (Fig. 20). Since this process is continuous, the oceanic lithosphere generally consists of well-defined uniform layers. This is significantly different from the structure of continental crust with its complex, long-lasting, and non-uniform history.

The asthenosphere consists on average of a garnet-bearing lherzolite as a specific type of peridotite (type locality at Lac de Lherz, French Pyrenees). It is mostly composed of olivine, a magnesium-silicate undersaturated in silica ($(Mg,Fe)_2[SiO_4]$) and with approximately 10 % of Mg replaced by Fe, and of two different pyroxenes, the orthopyroxene enstatite ($Mg_2[Si_2O_6]$) and the clinopyroxene diopside ($CaMg[Si_2O_6]$; Fig. 21). This primary composition, however, is modified when the lherzolitic peridotite ascends into the asthenospheric wedge below the mid-oceanic spreading center (Fig. 19).

The rise of peridotite causes an increase in partial melting because of pressure reduction where the amount of partial melting may increase from less than

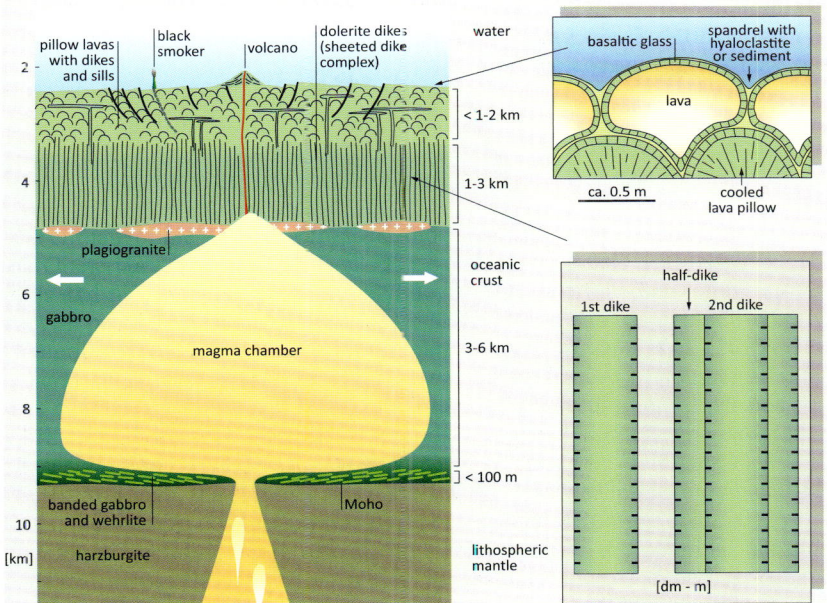

Fig. 20: Formation of oceanic crust from a magma chamber beneath the mid-ocean spreading center. Insets show formation of pillow basalts and dike-in-dike structures with half-dikes (modified after Frisch et al. 2011, Frisch & Meschede 2013).

5 % to partly more than 20 % directly below the mid-oceanic spreading center. The melt extracted from the lherzolite has a basaltic composition which is roughly similar to the composition of diopside. The enclosed molten blobs are of lower density which accelerates the rise of the peridotite. Continuing pressure-reduction increases the amount of basaltic melt until the diopside completely disappears from the peridotite. Therefore, the original lherzolitic peridotite changes into a harzburgite composition (type locality at Bad Harzburg, Germany) which is mainly composed of olivine and enstatite (mostly bronzite).

A simple formula for the extraction of basaltic melts from peridotite is expressed by

lherzolite = harzburgite + basaltic melt

where the basaltic melt in the magma chamber beneath the mid-oceanic spreading center is an extract of 10–25 % from the original lherzolitic melt. After solidification, gabbros formed from this melt are mainly composed of diopside and plagioclase, a calcium-sodium feldspar which is a mixed crystal

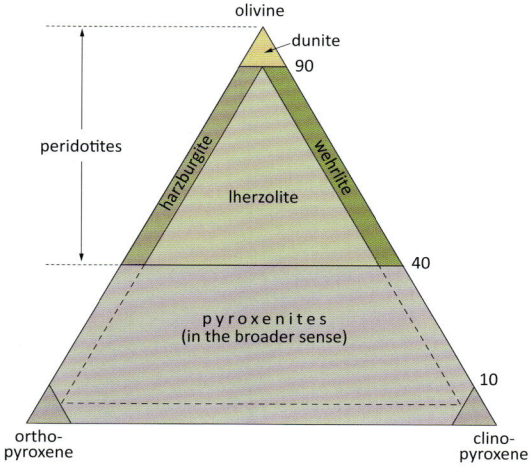

Fig. 21: Composition of ultramafic mantle rocks (numbers are in %).

between albite ($NaAlSi_3O_8$) and anorthite ($CaAl_2Si_2O_8$). Cumulates at the base of the gabbroic layer near the Moho (Fig. 20), however, may be composed of olivine and clinopyroxene, at some places forming accumulations of wehrlite (Fig. 21). Oceanic basalts formed at mid-oceanic spreading centers are named tholeiites (type locality at Tholey, Germany). They differ significantly from alkaline basalts formed at intraplate locations or calc-alkaline basalts formed above subduction zone systems.

Emplacement of the Samail Ophiolite

The Samail Ophiolite of Oman is characterised by its large surficial extent and the high quality of exposures under the arid climatic conditions of Oman. It is the largest coherent ophiolitic body and one of the best exposed ophiolites of the world, namely 500 km long, 50 to 100 km wide, and 15 km thick. The ophiolite was thrust as a large nappe onto the passive margin of the African Plate of which Arabia was part. The ophiolite nappe was initially thrust several hundred kilometers over oceanic crust that belonged to the other side of the spreading axis. It rapidly migrated southwestward and the entire complex was obducted onto northeastern Arabia at 80 Ma (Fig. 16). The calculated thrusting velocity was approximately 3 cm per year (Nicolas 1995, Frisch et. al. 2011, Frisch & Meschede 2013).

Furthermore, the Samail Ophiolite consists of two different types of magmatism: magmatic rocks formed at a spreading center (type 1) of the Neotethys Ocean from the Triassic to the Middle Cretaceous. A later episode is associated with the onset of subduction or may have been created by melting processes at the base of the overthrusting ophiolite unit (see below). During the spreading center phase mostly basaltic and gabbroic rocks of MORB geochemistry were formed in a classic oceanic crust sequence. During the later phase irregular magmatic bodies of basaltic to andesitic composition (type 2) intruded into the layered rocks of type 1 in a typical supra-subduction zone (SSZ) environment. The age of type 2 magmatism has been determined by U-Pb dating, with 95.3 million years as the best representative age (Goodenough et al. 2014). Therefore, caused by the thrusting over the existing ocean floor, the ophiolitic rocks are enriched in SSZ magmatic rocks.

When the ophiolite nappe was thrust over the continental margin, the forces of friction increased due to the buoyancy of the continental crust. Therefore, obduction ceased after the nappe was transported 100–200 km over continental crust. During early stages of thrusting, a several hundred meters thick metamorphic sole formed at the base of the ophiolite nappe in the rocks of the overridden oceanic crust below, which is a characteristic feature of Tethyan ophiolites according to Nicolas (1989). During obduction, heat from the hot base of the overthrusting ophiolite nappe is transferred downwards into the units below the thrust surface (Fig. 17). The heat induces metamorphism and is accompanied by deformation caused by the thrusting process. This style of metamorphic zone below the ophiolite nappe is called a metamorphic sole. It is characterised by metamorphism that is most intense at the tectonic boundary and rapidly decreases with depth because the generation of heat persisted for only a short time span. The metamorphic sole of the Samail Ophiolite is composed of amphibolites and quartzites derived from overthrust basalts and radiolarites, respectively. Peridotites of the ophiolite attained mylonitic deformation at about 1000 °C. At the contact between these mylonites and the underlying deformed units, strong foliation occurs, induced by the thrust movement along the nearly horizontal thrust plane. Basalts of the subducted ocean floor were heated to 900 °C and, besides the metamorphic alteration to amphibolite, partly melted along with overlying sedimentary rocks. These melts may have created the andesitic volcanism in the overthrusting unit, which has the characteristics of island-arc volcanism (Boudier et al. 1988). Other authors interpret this unit as a result of a supra-subduction volcanism (e.g., Gray and Gregory 2003). During continued thrusting of the ophiolite nappe over the ocean floor, the base rapidly cooled and progressively lower temperature metamorphism overprinted the underlying basalts (e.g., greenschists in the temperature range between 500 °C and 300 °C).

The initially flat and thin lithospheric nappe has been detached and thrust over a sole of wet and heated oceanic crust, partially with its overlying sediments (Fig. 17). Nicolas (2016) describes this as the "ironing effect" of the ophiolite nappe which is only possible if the detachment occurred near the ridge axis. New and precise age datings in the Samail Ophiolite (Warren et al. 2005, Rioux et al. 2013) indicate that only a very short age span of about 300,000 years lies between the latest magmatic events related to the activity at the spreading center and melting at the metamorphic sole.

Based on geochemical analyses of basaltic rocks, the Samail Ophiolite has been described as formed in a back-arc environment (Lippard et al. 1986). However, in the Hawasina basin, which was overridden by the ophiolite nappe, arc-related volcanics as expected in such a tectonic environment, have never been described. The Samail Ophiolite is, therefore, interpreted as a MOR ophiolite which was obducted onto a continental margin (Nicolas 2016).

Excursus II: Snowball Earth – the largest glaciation of the Earth's history

Prolonged and intense glaciation phases occurred several times during the Proterozoic. They were characterised by extensive ice sheets, which in some cases covered even entire continents. In some places evidence for these glaciations are preserved, such as in the center of the Oman Mountains (cf. EP 55, EP 62). It was first assumed in the 1960s that one or more of these glaciations affected the whole Earth and that a closed ice cover stretched from pole to pole (e.g., Harland 1964). Kirschvink (1992) was the first who used the term "Snowball Earth" for this hypothesis.

The model of a planet-wide Proterozoic glaciation, however, is still under discussion. A major point of controversy is the question how life forms which already existed at the time, could have survived under a closed ice cover (Gaidos et al. 1999, McKay 2000). It is also unclear how the very strong albedo effect of such a glaciation allowed the transition to normal, ice-free conditions. Ice has a high albedo, meaning that most sunlight and therefore solar energy, hitting the surface bounces back towards space. It is, nevertheless, assumed today that the Earth has been in the Snowball Earth-state at least three times: the first glaciation was the Makganyene glaciation in the Early Proterozoic, about 2.3 billion years ago, followed by the Sturtian, around 720–660 million years ago and then the Marinoan glaciations of the late Proterozoic, between 650 to 635 million years ago (Arnaud et al. 2011).

Excursus II: Snowball Earth – the largest glaciation

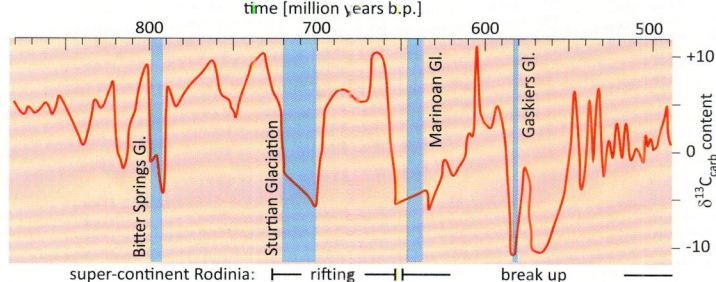

Fig. 22: δ¹³C ratio in calcareous sediments of the Late Proterozoic (simplified after Halverson et al. 2005).

The Snowball Earth model describes a state of the Earth with a global average temperature of about –50 °C. The mean-annual surface temperature around the equator was around –20 °C and resembles the present temperatures in high polar regions. H_2O was transported from areas with strong sublimation i.e. the direct transition from the solid to the gaseous phase, into areas where condensation predominated. This resulted in differences in regional glacial mass balances which were balanced by ice movements. Glacial sediments such as tillites and other diamictites, thick glaciomarine sequences with dropstones etc. developed as a result of these movements. Fortunately they have been preserved as evidence of these glaciations (Pierrehumbert et al. 2011).

Based on paleomagnetic studies and the global distribution of glacial sediments, it can be deduced that at least parts of the glacial sediments of the Late Proterozoic were deposited near the equator. This led to the assumption that the Earth must have been completely covered with ice, even up to the equator and including the oceans. Other authors suggest a belt with open water around the equator (Pierrehumbert et al. 2011). The extreme glaciation is also documented by the carbon isotope ratio ($δ^{13}$ C) which indicates the intensity of biological activity. Late Proterozoic glacial sediments are characterised by a particular low $δ^{13}$C ratio (Fig. 22) indicating an extremely low level of biological activity (Hoffmann et al. 1998).

The strong cooling of the Earth was caused by a significant reduction in greenhouse gases; mainly carbon dioxide (CO_2) and methane (CH_4). At the end of the Proterozoic, the continental land masses concentrated around the equator in the supercontinent Rodinia. A large landmass around the equator, however, resulted in intense weathering of the rocks cropping out at the surface. Weathering of limestone and silicate rocks led to the removal of CO_2 from the atmosphere by the formation of hydrogen carbonate (HCO_3^-). Calcium ions from silicate minerals, e.g., from plagioclase (Ca-feldspar) or augite (Ca-containing pyroxene) were dissolved with the aid of hydrogen carbonate removed by water.

Extensive land surfaces around the equator favor the cooling of the Earth, because, in contrast to the oceans, land surfaces have a higher albedo. Therefore, just where the sunlight is energetically most intense due to the sun's position, the Earth can absorb less heat. Moreover, the solar radiation in the Proterozoic was about 6 % lower than today and large volcanic eruptions carried huge amounts of ash into the atmosphere. Feedback effects such as the increasing albedo led to global glaciation because of their multiplier effect. At the same time, the albedo effect is one of the main arguments against the Snowball Earth hypothesis, because, according to calculations, a return to non-glacial conditions would be virtually impossible – and the Earth would be left permanently in a Snowball Earth condition. Since this was not the case, a mechanism is needed that results in global ice melt and termination of the state of Snowball Earth, regardless of the quantity of the supplied solar energy.

Possible causes are plate tectonic movements and processes that are not interrupted even in case of an extensive ice cover and strongly reduced biological activity. The geochemical carbon cycle continues even with a complete ice cover. Carbon in the form of $CaCO_3$ and hydrocarbons, i.e. organic components in sediments, are subducted and affected by higher pressure and temperature. $CaCO_3$ reacts, for instance, with SiO_2 under separation of CO_2 to silicate minerals (e.g., wollastonite: $3\ CaCO_3 + 3\ SiO_2 = Ca_3(Si_3O_9) + 3\ CO_2$) and organic components are converted into methane. Volcanoes above subduction zones release the separated carbon dioxide and methane in the form of volcanic gas.

Therefore, carbon dioxide reached the atmosphere in large quantities also during the Snowball Earth-state. However, since weathering on the surface of the Earth was mostly prevented by the ice cover, there was no possibility to bind the released carbon dioxide, for example, by silicate weathering (Kirschvink 1992). As a result the greenhouse gas was again gradually enriched in the atmosphere.

At the end of the global glaciations in the Late Proterozoic, the atmospheric CO_2 concentration reached about 10 %, a value which is by order of magnitude higher than the current CO_2 content which is only 0.04 %. At a certain CO_2 level the climate system reaches the tipping point which results in a rapid warming and the development of an extremely agitated super-greenhouse climate (Allen & Etienne 2008). To accumulate enough CO_2 in the atmosphere for the Earth to be able to leave the Snowball state, a period of at least 5 to 10 million years is required.

Excursus II: Snowball Earth – the largest glaciation

At the tipping point, rapid warming melted down the entire ice sheet within a short period of time, which probably lasted only a few thousand years (Fig. 23). As a result, the sea level rose rapidly by several hundred meters, perhaps even more than one kilometer. The release of methane which is a much more powerful greenhouse gas than CO_2, also contributed to the rapid warming (Kennedy et al. 2001). Marine gas hydrates are stable in a solid form only at low temperatures. They decompose at warm temperatures under normal pressure conditions and large quantities of methane can be released in a short time.

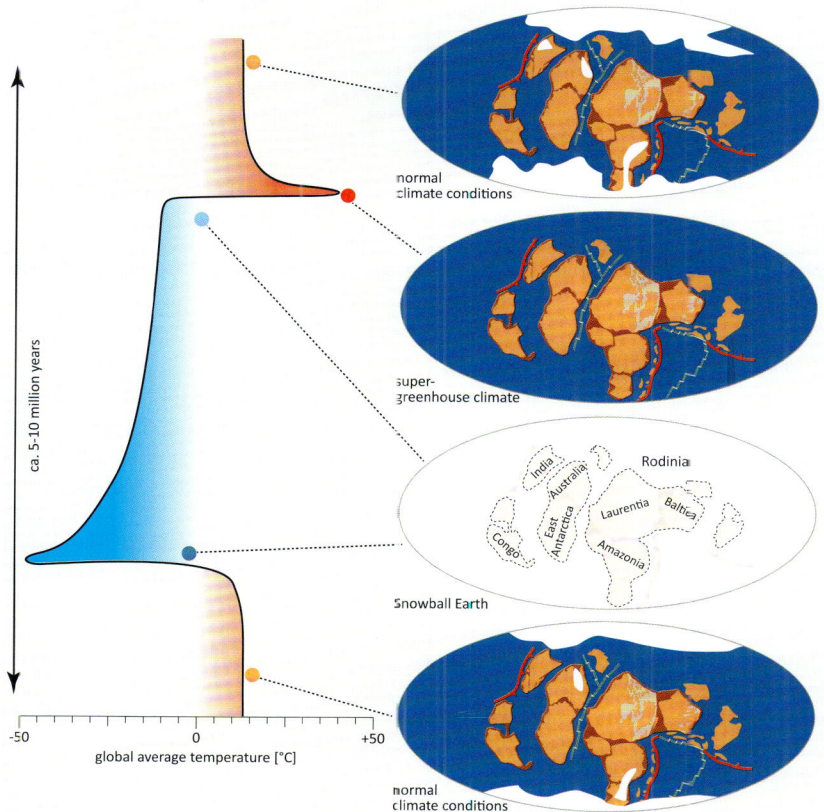

Fig. 23: Development of the Snowball Earth-stage and the temperature at the end of the Sturtian glaciation in the Late Proterozoic about 660 million years ago (simplified and modified after Pierrehumbert 2002, Fairchild & Kennedy 2007).

Fig. 24: Late Proterozoic glacial diamictites of the Fiq formation overlain by cap dolomites of the Hadash formation in the Wadi Haslan, Jebel Akhdar Mountains. The diamictites belong to the Sturtian glaciation phase. For location see EP 62 (Figs. 145, 158).

The melting of the ice also restarted the chemical weathering processes so that atmospheric CO_2 could bond again. CO_2 was washed out of the atmosphere due to heavy rain and fell down as acid rain, accelerating the weathering on the surface of the land. At the same time abundant calcareous sediments formed in the warming seawater. This resulted in a sedimentary sequence typical for the end of the Proterozoic glaciations.

In the present time, dolomites and calcareous sediments of up to 30 m thick, known as cap dolomites, are found directly on top of the glacial sediments worldwide (Fig. 24). They differ from conventional carbonates and have unusual properties such as giant wave ripples, microbial mats, vertical tubular structures or barite ($BaSO_4$), formed in an early stage of diagenesis. They originated on carbonate platforms, shelfs or submarine slopes and even at locations which usually are not typical for the deposition of carbonates. Most of the cap dolomites are of transgressive origin thus indicating the flooding of the continental margins due to the melting ice.

The glaciations of the Snowball Earth are also related to the increase of the oxygen content in the atmosphere at those times (Fig. 25). A first sharp increase in the oxygen content occurred after the Early Proterozoic Makganyene glaciation, about 2.2 to 2.4 billion years ago, with an increase of the oxygen

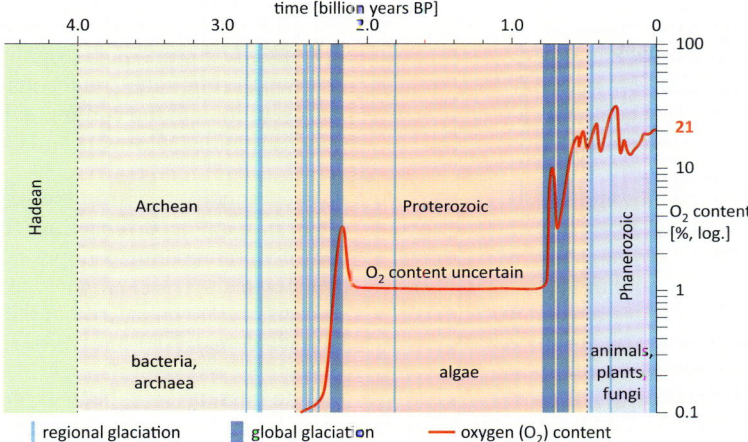

Fig. 25: Glaciation periods and the evolution of the oxygen content during the Earth's history (modified after Goldblatt et al. 2006, Berner 2006, Harada et al. 2015).

content from a value below 0.1 % to about 1 %. In the late Proterozoic, the oxygen content increased after two Snowball Earth glaciations and reached a level of about 15 %. It is assumed that after the Marinoan glaciation a large supply of nutrients led to a widespread development of organic rich sediments. The oxidation of the organic components were prevented by their rapid sedimentation and the atmospheric oxygen was not consumed, which as a consequence was enriched in the atmosphere. This increased oxygen content allowed the rapid development of the living environment after the glaciations with complex multicellular organisms that needed significantly more energy. This energy was supported by the offered oxygen. Thus, it is postulated that the Snowball Earth-states probably played a significant role in the development of life on Earth.

5.4 Main mineral resources of Oman

Oman's natural mineral resources are diverse and widely spread. Many minerals are present in commercial quantities and have been exploited for thousands of years. Today, Oman produces quantities of both metallic and non-metallic minerals. The most common metallic minerals are copper, chromite, manganese and gold, whereas the most abundant non-metallic ones are limestone, gypsum, dolomite, laterite and coal. Different types of aggregates, crushed sands and gravels, are produced from fluvial deposits along the main wadi streams. Metallic minerals are predominantly found in the ophiolitic sequence, which was emplaced in northern Oman during Late Cretaceous (see Excursus I). The non-metallic minerals are present in both the continental and oceanic sediments that were deposited mainly during the Mesozoic and Cenozoic eras.

The metallic mineral potential in Oman is essentially found in the Samail Ophiolite (Le Metour et al. 1995). Hundreds of mineral occurrences are found in northern Oman, some of which are of good economic value. Copper occurrences are found in the upper part of the sequence within massive chalcopyrite bodies that were formed by the hydrothermal processes, essentially at the end of submarine volcanic cycles during Late Cretaceous (cf. EP 91–EP 93). These were later deposited on the sea floor.

The chalcopyrite bodies or massive sulphide gossans have been exploited for copper since at least 5,000 BP. Gossan is an intensely weathered or decomposed rock, usually the exposed part of an ore deposit. They appear as minor topographical highs, light-red or brownish-red in colour, with commonly elliptical shapes of about 30 to 50 metres in length. Most of these are present in northern Oman, close to the wilayats of Sohar and Shinas, where the highest and thickest volcanic successions of ophiolites occur. The legendary land of Magan, as mentioned in the Sumerian texts, is identified as the land from where cargoes of copper and diorite where exported. Tens of ancient copper mines are found in north Oman, particularly in Wadi al Jizzi in Sohar.

Chromite accumulations occur within the upper part of the ophiolitic mantle sequence. They mostly form within the dunite and harzburgite bodies, within a transition zone of mixed dunite and harzburgite, normally less than 1 km below the line of Moho, that separates the crustal rocks from the mantle sequence (Qidwai 2004, Le Metour et al. 1995). Al Azri (1987) classified the chromite pods in Oman into three groups: type A corresponds to tabular and lenses within the mantle peridotite, type B pods form in the upper part of the harzburgite and tend to be very deformed, and type C are comprised of stratiform chromite deposits that form a pile of alternating chromite beds and dunite beds. Compared to other ophiolite occurrences, e.g. in Greece and the Philippines, the Samail Ophiolite has a lower chromite content. Nicolas & Al Azri (1991) explain this by suggesting that the ophiolite in Oman was formed in a fast spreading ridge. As a result, the rapid spreading did not allow for low temperature conditions, which are necessary for the crystallisation of chromite.

Gold, silver, lead, manganese, iron ore, zinc, platinum and nickel are other mineral occurrences found within the ophiolite complex and the deep oceanic sediments. Overall, small gossans or associated grains of these minerals are found, particularly within the volcanic rocks of the crustal ophiolite or in the

Hawasina rocks. The reserves of these minerals in Oman are generally low, but the possibility of discovering new economical potentials remains high. The manganese deposits near the wilayat of Sur can be upgraded to produce a range of manganese chemicals. Furthermore, occurrences of iron, manganese and uranium have been reported in Cretaceous and Tertiary sediments in the Dhofar governorates (Le Metour et al. 1995). The Tertiary rocks of Dhofar also have potential sites to exploit phosphate minerals. These mainly occur as small lenses in the shale deposits of this time period. However, the content of phosphate minerals and their lateral extension is generally limited.

Numerous occurrences and extensive deposits of evaporite minerals and carbonate rocks are found in south, central and north Oman. These are mainly exploited as construction materials and ornamental stones. Gypsum is abundant in the Tertiary deposits in the governorates of Dhofar and Al Wusta, in south and central Oman. Gypsum deposits are also present in north Oman (Al Buraymi governorate) as small beds. Oman is currently considered as one of the leading countries in the region in producing good-quality gypsum. The mineral is mainly used and processed in cement factories in Oman and the UAE.

Limestone and marble-like limestone are also used in cement factories and as construction bricks or ornamental stones. Oman's marble-like limestone has a mostly pale yellow or whitish grey colour. These are either produced from Tertiary limestone or more commonly from olistoliths which are exotic blocks of shallow marine limestone that were transported by submarine gravity and included within deep oceanic sediments, or Mesozoic atolls that once formed ring-shaped coral reefs in the Neotethys Ocean and were later emplaced in Oman during Late Cretaceous by the obduction process. Overall, the limestone in Oman is of excellent quality and is more or less easily accessible (Qidwai 2004). Many exotic blocks of limestone can be sited along the road from Muscat to Nizwa and from Nizwa to Ibri. Some of these are mined for commercial marble. Chalky limestone is also available in northern and southern Oman. These rocks are suitable for low grade filler. Dolomites in the Tertiary deposits of northeastern Oman can be considered as raw material for the glass industry.

Oman also has significant resources of a number of clay minerals, coal, aggregates and sands, including kaolinite, smectite, alluvial sediments and silica sands. The most known kaolinite deposits are located in central Oman. However, these deposits have a high percentage of iron oxides; therefore their use in manufacturing is limited. Likewise, the quality of potential sources for bentonite and smectite clays in Oman is generally not high, hence limiting their economic potential. However, the deposits of clay in the Late-Cretaceous Fiqa Formation in northern Oman can have good applications in the manufacturing of bricks and tiles, although additional studies are required to evaluate the quality and possible purification (Le Metour et al. 1995). Deposits of clay minerals are also found in a number of locations in the northeastern part of Oman, around the wilayat of Sur. In the same area, more specifically Wadi Msawi and Wadi Fisaw, multi-seam deposits of coal are formed in the Tertiary deposits, alternating with sand and shale beds. The coal layers form together good reserves of coal and lignite. The government has recently considered developing these reserves as a fuel for electrical power generation.

Reserves of aggregates and silt are exceptionally high in Oman. A variety of good-quality options is available because of the different setting of recent wadi-alluvium deposits, with various grain sizes, shapes and distributions, which can serve the local needs for decades. Many exploited sites of aggregates are currently located around industrial areas in the main cities as the country aims to outline the areas where aggregates are recovered from. Potential exports of aggregates to neighbouring countries remain very high as Oman has large resources that can meet the needs of the Gulf countries.

Oman could also be a potential producer of quartzite and silica sand. A number of sites are found across Oman where these minerals can be exploited. The most well-known sites are the quartzite deposits of the Early-Palaeozoic Amdeh formation in northern Oman and the Abu Tan Sand in the Late-Cretaceous Samhan formation in central Oman. Glass sand could be easily obtained from the latter site, however the site is located in the Oryx Reserve, hence its exploitation is prohibited.

The recent establishment of the Public Authority of Mining and the emergence of new mining laws and regulations are expected to promote the exploration and production of minerals in Oman. The authority also takes a leading role in producing and providing geological maps, studies, surveys and laboratory services in Oman. The new governmental initiatives will hopefully increase the contribution from the mining sector in the annual GDP of the country and support the sustainability of business in the long term. Exploitation of minerals in Oman requires an exploration program and environmental license from the Ministry of Environment.

5.5 History of oil and gas exploration and production in Oman

After more than 35 years of surveying and exploration, from the end of 1920s to the beginning of 1960s, sequential attempts to discover commercial quantities of oil resulted in the birth of a new hydrocarbon-producing state. This oil strike was later to support transporting Oman into a new renaissance period.

The onshore and offshore parts of Oman are currently divided into about 50 concession blocks and there are about 15 operating companies that explore and produce hydrocarbons in Oman. Oman's hydrocarbon formed from remnants of organisms, most of which had primitive single-celled forms and lived more than 500 million years ago. These organisms floated in the oceans and after death were deposited under conditions absent of oxygen; a necessary precondition for hydrocarbons to form. Lijmbach et al. (1981) conclude that all organic matter found in the source rocks is derived from marine organisms, as no land-plant material could be identified.

Most of the crude oils in Oman are characterised as light oils with low-sulphur and low-wax content. Exceptions are the crude oils in the south. Grant-

Fig. 26: Location of major oil and gas fields in Oman as well as related infrastructure.

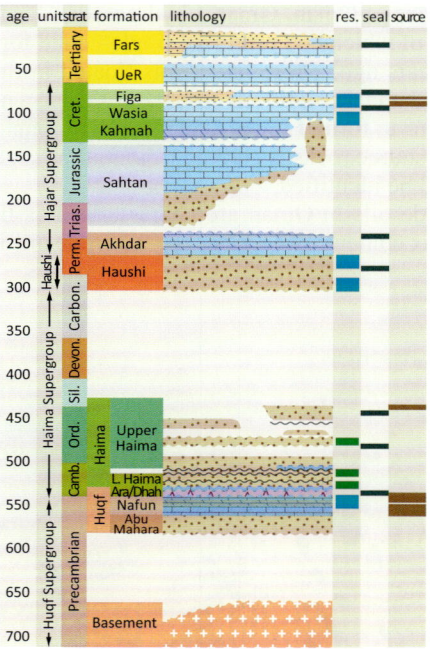

Fig. 27: Simplified stratigraphy of Oman highlighting the main reservoirs (blue – oil, green – gas), seals and source rocks (modified after Terken et al. 2001).

ham et al. (1988) discriminate five distinct oil families in Oman, depending on the source rocks and based on differences in molecular fingerprints (see also Terken et al. 2001). In fact, some of the source rocks of hydrocarbons in Oman are among the oldest in the world, namely the oil deposits located in the southern Oman salt basin (Grosjean et al. 2009). The burial history and tectonic forces have converted the organic matter of these organisms into a mixture of complex compounds of hydrogen and carbon (hydrocarbons) under conditions of high temperature and pressure.

The hydrocarbons produced in Oman have a wide range of densities and chemical signatures. The oil is mainly produced from Mesozoic carbonate reservoirs in northern Oman and Palaeozoic sandstone reservoirs in southern Oman (Figs. 26, 27). Hydrocarbons are also produced in lesser quantities from carbonate and clastic Precambrian reservoirs in southern and central Oman.

Glennie (2005) subdivided 3 main sequences (Fig. 27):
1. Precambrian to Lower Palaeozoic Huqf and Haima Supergroups within the southern Oman salt basin with the Ara Salt as a seal
2. Permo-Triassic Haushi and Hajar groups, sealed by intraformational shales and
3. Jurassic to Cretaceous carbonates of Oman's interior, sealed by Natih and Fiqa shales (see also Clarke 1988, Forbes et al. 2010).

Oil

The story of exploring and producing oil in Oman is full of setbacks and excitements. First serious attempts to understand the geological settings of Oman were conducted by D'Arcy Exploration in the mid-1920s (Fig. 28). The company obtained a two-year license, subject to renewal, from Sultan Said bin

Fig. 28: The geological survey team of D'Arcy Exploration in 1925/1926, led by second from right George Lee (Source: Royal Society for Asian Affairs).

Fig. 29: René Wetzel (left) and Mike Morton who surveyed Dhofar in 1948 (Source: Mike Morton).

Taimur. The survey team was led by George Lee and operated in volatile conditions in the interior part of the country, yet it also managed to cross a great part of Western Al Hajar Mountains using camels. The initial findings were not encouraging, particularly with the formidable natural barriers, although geological structures where oil could be trapped were found.

In 1937, Iraq Petroleum Company (IPC), which represented a consortium of oil companies including Shell, BP and ExxonMobil, reached an agreement with Sultan Said bin Taimur to explore the northern and southern parts of Oman for oil. For this purpose, they created a new company called Petroleum Development (Oman and Dhofar) Ltd. The company sent two renowned geologists, Mike Morton and René Wetzel, to survey the Dhofar area (Fig. 29). After six weeks of surveying, they concluded that the Dhofar area is geologically not viable for oil exploration. IPC then decided to relinquish its petroleum concession in southern Oman in 1950 and the company was renamed Petroleum Development (Oman), simply referred to as PDO until today.

As a result and inspired by the successful discoveries made by American companies in other parts of the Middle East, Sultan Said bin Taimur granted the concession to an American company, Dhofar-Cities Service Petroleum Corporation. Dhofar-Cities Service expended around $50 million and drilled about 30 wells in Dhofar, among which were Dauka-1, the first petroleum well drilled in Oman, and three other wells in the Marmul area. These wells were drilled between 1955 and 1958. Despite finding hydrocarbon, the oil was too heavy to flow to the surface, hence it was not feasible to exploit

Fig. 30: A photo of Fahud-1, taken in the 1950s, in Jebel Fahud (Source unknown).

commercially. Eventually, the American company relinquished the Dhofar block in 1967. The block was re-acquired by Petroleum Development (Oman) in 1969.

The Iraq Petroleum Company (IPC) reconfirmed its agreement with Sultan Said bin Taimur and commenced its activities in northern Oman in 1948. After establishing a base in Ad Duqm, the company defined Jebel Fahud as its first drilling target in 1954, following an identification of this low-lying structure from an airline flight. However, the drilling of Fahud-1 exploration well in 1956 (Figs. 30, 31) ended up to be one of the unluckiest misses of oil reservoirs in the history of hydrocarbon industry. The well penetrated the reservoir only a few tens of meters from what would have been a big discovery (Searle 2014, Morton 2006). IPC drilled 3 other petroleum wells in northern Oman in the late 1950s, but they were all disappointing. Following the instabilities in the area and the high supply of crude oil in the world at that time, three parties of IPC decided to withdraw from the consortium, leaving behind only Shell and Partex as partners in PDO, with 85 % and 15 % shares respectively.

With the onset of the 1960s, the oil prices had already recovered as OPEC was formed and Oman was becoming more stable for oil companies to operate in. Furthermore, seismic survey technology had improved significantly and a new exploration era had just started as better datasets were acquired. PDO started a new drilling programme between 1962 and 1964. The programme led to the discovery of three big oil fields in northern Oman: Yibal, Natih and Fahud, which was previously missed (Fig. 26). The company extended a main

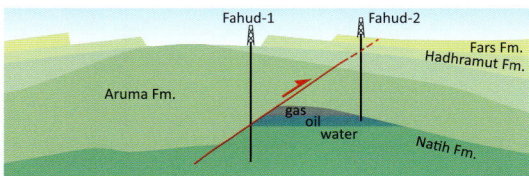

Fig. 31: Structural setting of Fahud structure, showing Fahud-1 and Fahud-2 wells. Fahud-1 missed the hydrocarbon reservoirs in the 1950s by only a few tens of meters. The giant reservoir was discovered in 1964 by Fahud-2.

pipeline of 279 km in length from these fields to the new oil exporting port in Mina Al Fahal in Muscat. The first export of oil occurred in 27 July 1967 in a shipment of 543,800 barrels that was sold for $1.42 a barrel and loaded into a Japanese tanker. The oil was pumped through the pipeline to the highest point in the wilayat of Izki at an altitude of about 650 meters and then naturally flowed by gravity to Muscat. In 1974, the Government of Oman acquired a 60 % stake in PDO.

Big discoveries in northern and southern Oman during the seventies contributed to the entry of a new era of professionalism in oil industry, which coincided with high oil prices. Among the fields discovered during that time are Saih Nihayda, Saih Rawl and Qarn Alam (Fig. 26). The oil production rate increased to 341 thousand barrels per day in 1975. In the late 1970s, the focus of exploration moved to southern Oman as this area became more stable. Many large fields were discovered along the eastern border of southern Oman during that period, including Amin, Nimr, Mukhaizna and Rima (Fig. 26). The concession area of PDO was reduced in the 1970s and 1990s to reach about 113,000 km^2, therefore allowing more companies to explore and deploy technologies in a quest for exploration and production of hydrocarbon resources in Oman.

During the 1980s and 1990s, the oil and gas industry in Oman became more prosperous and mature. A new record of production was achieved in 1984 with 400 thousand barrels per day and reserves amounted to 8.3 billion barrels. In 1986, oil prices collapsed, therefore necessitating the Sultanate to increase production and lower the expenses. This goal was supported by the wide use of 3D seismic survey at that time and the appearance of horizontal wells, which increased production by 2 to 4 times compared to vertical wells. Oman has since then been recognised as a pioneer in the use of these technologies to enhance hydrocarbon production.

In the mid-1990s, the reserves and production increased steadily, reaching 761 and 840 thousand barrels a day in 1994 and 2000 respectively. However, the production rates from the main oil fields dropped dramatically at the begin-

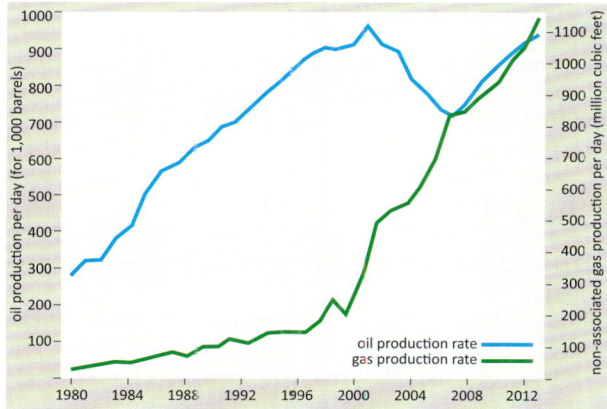

Fig. 32: Oil and gas production rates in Oman from 1980 to 2013 (data source: National Centre for Statistics and Information 2014).

ning of the 2000s (Fig. 32), creating an urgent incentive to increase subsurface studies to understand the complexity of oil reservoirs and identify the best methods to develop them and sustain production.

In the wake of the new subsurface challenges in the 2000s, Oman has taken thought-provoking steps towards ambitious plans to use enhanced-oil recovery methods to extract hydrocarbons from reservoirs. These include strategies other than the traditional recovery methods that depend on natural reservoir depletion or water-flooding the reservoirs. Examples of these methods include flooding the reservoirs with hot gas or steam, injecting chemical solvents or using different polymers. However, these techniques need a lot of investment and require proper understanding of reservoir complexities. For this purpose, study teams have been deployed to produce field development plans that can be used as roadmaps to enhance oil recovery.

In 2004, the government also extended its agreement with the other PDO's shareholders for another period of 40 years, ending in 2040. By 2005, the oil production of Oman dropped to less than 700,000 barrels per day and the need for applying enhanced-recovery methods for extracting oil increased. These methods are currently applied in large fields like Mukhaizana, Qan Alam, Marmul and Harweel (Fig. 26). In July 2015, Oman crude and condensate production exceeded 1 million barrels a day for the first time. See Clark (2007) for a full account of the history of hydrocarbon exploration in Oman.

Gas

Two types of natural gas are found in a number of fields across Oman. The first is associated with oil reservoirs (associated gas) and the second is non-associated gas. The first type was discovered in the 1960s in fields like Yibal and Natih, and started to receive more attention in the 1980s as the demand for liquefied gas around the globe was increasing, mainly for power plants and industrial areas. Under the auspices of His Majesty Sultan Qaboos bin Said, the government opened a large gas-treatment plant in Yibal Field in 1978. Since 1981, the treated gas has been transported to Sohar through a new pipeline of 226 km in length.

The second type was primarily discovered in the 1990s in large fields in central Oman (e.g. Saih Rawl, Saih Nihaydah and Barik fields; Fig. 26). Deep wells, some of which are more than 6000 m deep, penetrated the gas reservoirs from the 1990s onward. The government developed large gas fields in central Oman and transported the gas through a pipeline of 350 km in length to Qalhat, near the wilayat of Sur. Here the Liquefied Natural Gas (LNG) plant was established. The first shipment of liquefied gas was to South Korea in 2000. The LNG plant can produce about 10 million tons of liquefied gas per year. Another gas treating plant was launched in Saih Nihaydah Field in 2005.

The economy of Oman is heavily dependent upon the oil and gas industry. In 2014 about 50 % of the total GDP is contributed by this sector, and 84 % of government revenues are derived from the hydrocarbon sector (Central Bank of Oman 2015). It is predicted that the country will continue producing its reserves of hydrocarbons for at least 50 years to come. Oman continues to develop its human resources to meet the high demand of qualified geologists and engineers in this business, particularly as challenges to extract more hydrocarbons from the reservoirs increase with time.

6 Field sites

EP 01

Muscat Harbour
Note: Sturdy shoes are recommended.
Topics: Ultramafic rocks, Samail Ophiolite, magnesite, scenic walk
Locations: 1a: UTM 40 Q 660682 2612907 / N 23°37'07" E 58°34'30"
            1b: UTM 40 Q 660579 2612708 / N 23°37'01" E 58°34'26"
            1c: UTM 40 Q 660266 2612444 / N 23°36'52" E 58°34'15"

Rating: ☺☺

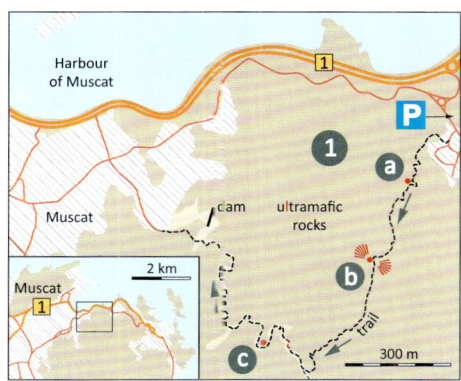

Fig. 33: Location map.

Approach: Follow the coastal main road #1 to the harbour in the old town of the city of Muscat and continue straight down the corniche with the sea on your left to a roundabout. Bear left at the roundabout into Riyam Street. There is a carpark on the left, about 200 m on from the roundabout. A steep, narrow trail starts up the hill on the other side of Riyam Street to a pass which is approximately 120 m above sea level. The downhill trail proceeds through a narrow but walkable river gorge (around location 1c, Fig. 33).

The complete loop of the trail starting at the car park is about 3.7 km. Around 2 km of this is a rocky footpath and takes around 1½ to 2 hours to complete, depending on your pace. A few days after rain is a particularly good time to visit.

The trail passes through a series of well exposed ultramafic rocks which belong to the Samail Ophiolite nappe. A number of magnesite dikes of decimeter-thickness can be

Fig. 34: Magnesite dike of cryptocrystalline magnesite in ultramafic rocks of the Samail Ophiolite (location 1a in Fig. 33).

Fig. 35: Layered ultramafic rocks near the old town of Muscat. This is the view from location 1b in Fig. 33 towards the north.

observed at location 1a (Fig. 34). Magnesite ($MgCO_3$) develops as a metasomatic alteration product of metamorphosed ultramafic rocks. Serpentinised olivine rich peridotite is altered to talc and magnesite when carbon dioxide, which is a common component of metamorphic fluids, is available ($2Mg_3Si_2O_5(OH)_4 + 3CO_2 \rightarrow Mg_3Si_4O_{10}(OH)_2 + 3MgCO_3 + H_2O$).

There are a number of scenic viewing points along the trail, e.g., at location 1b (Fig. 33) near the pass, where the layering of the ultramafic rocks is clearly visible (Fig. 35). The dark brown layers are composed of harzburgite, whereas the lighter yellowish brown layers consist of dunite containing chrome-spinel (Le Metour et al. 1986a). The layers have variable thicknesses from centimeters to several meters and can be studied in detail in the river gorge, as at location 1c, Fig. 33).

EP 02

Viewpoint into Wadi al Kabir, south-west of Muscat
Topics: Ultramafic rocks, Samail Ophiolite, Tertiary unconformity, Triassic dolomites, scenic view
Location: UTM 40 Q 660197 2606808 / N 23°33'49" E 58°34'10"

Rating: ☺☺

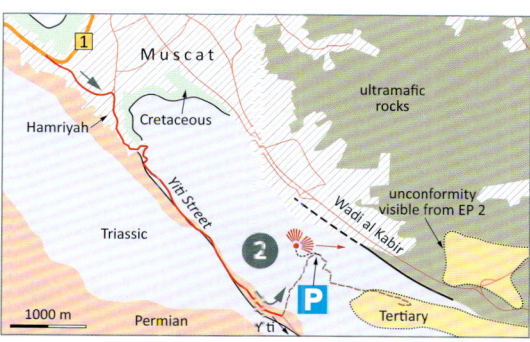

Fig. 36: Location map.

Approach: Drive up the winding road known as Yiti Street from Hamriyah, the urban district of old Muscat to Yiti. The view as you look back on the white houses of Hamriyah located on the valley floor is particularly picturesque. Turn left (UTM 40 Q 660041 2605934 / N 23°33'21" E 58°34'05") 3.2 km after leaving the last house. Take the graded road as far as possible and park. The last part of this 1000 m long access route may not be accessible by car. There is a paved walkway on top of the mountain and a nice viewpoint 200 m west of the road.

Fig. 37: Panoramic overview into Wadi al Kabir from the view point. Direction of view is towards the northeast (see Fig. 36). The hills are made up of ophiolitic rocks, overlain by Eocene marl (cf. Fig. 38).

This stop is located on a mountain overlooking Wadi al Kabir, located some 200 m below (Fig. 37). The Arabic word "kabir" means large. The wadi follows a major fault which strikes NW-SE and separates Triassic dolomites of the Mahil formation from peridoditic rocks of the Samail Ophiolite sequence. There is an ongoing discussion whether or not the fault is active, as the coastal configuration in this area indicates ongoing subsidence (Hoffmann et al. 2013a). The Triassic sequence is thrust onto Cretaceous sediments of the Muti formation which represents the top of the autochthonous

Fig. 38: Par-autochthonous Eocene marl unconformably overlying peridotitic rocks of the Samail Ophiolite nappe visible from the view point of EP 2 (see Fig. 36).

unit of the Arabian plate margin. The ultramafic rocks across the fault, (view towards NE) are mostly composed of dark harzburgite with some layers of lighter dunite. They form a rugged relief and are unconformably overlain by par-autochthonous Eocene strata (yellow marl). The contact between the Cretaceous igneous rocks of the ophiolite sequence and the Eocene marls is developed as a nonconformity (Fig. 38). The landscape characteristics are subject to change with ongoing land clearance in the area. The valley floors are widened to accommodate more commercial buildings.

EP 03

Drowned valley of Yenkit
Note: Check the tide table in advance as this site is only accessible at low tide.
Topics: Folded Permian dolomite, scenic view, wreck of an Omani dhow
Location: UTM 40 Q 674090 2601850 / N 23°30'15" E 58°44'45"

Rating: ☺

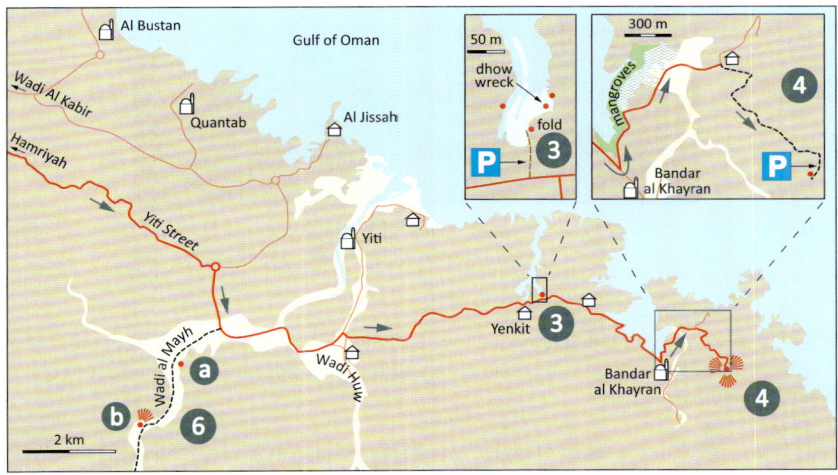

Fig. 39: Location map.

Approach: Follow the narrow road known as Yiti Street from Hamriyah, the urban district of old Muscat, to Yiti. Turn right in Wadi Huw (UTM 40 Q 660041 2605934 / N 23°33'21" E 58°34'05") following the signs to As Sifah. Continue along this road for about 5 km to Bandar al Khayran. You arrive at an easy access to the drowned valley of Yenkit on the left. There is a carpark near the road. About 100 m from the road, there is

Fig. 40: Coastal notch in the drowned valley of Yenkit. Folded Permian limestones and calc-schists are exposed with living oysters at the base, visible during low tide.

a wreck of a typical wooden Omani dhow, a traditional sailing vessel used in the Indian Ocean region. The valley is accessible only during low tide, so it is important to check the tides in advance.

The valley is cut into a series of uppermost Permian limestone and mudstone of the Saiq formation. They are strongly affected by folding and show strong schistosity which results partly in layers of calc-schists. A south-east vergent horizontal fold is exposed in the thinly bedded limestone on the eastern side of the valley. The valley belongs to a series of drowned valleys in the area south-east of Muscat, indicating a recently active subsidence movement in this region. During low tide a well-exposed recently formed coastal notch with oyster shells at the base can be observed (Fig. 40).

EP 04

Viewpoint Bandar al Khayran
Location: UTM 40 Q 678318 2600394 / N 23°30'14" E 58°44'46"

Rating: ☺☺

For location map see EP 03, Fig. 39.

Approach: The road to Bandar al Khayran goes through Yiti, which can either be reached via Hamriyah (Muscat; via Yiti Street) or through Qantab and Al Jissah. The road from Yiti to Bandar al Khayran is relatively narrow and winds through different gorges. In Bandar al Khayran, turn left and continue driving around the small mangrove swamp at the lagoon. Near the end of the paved road, turn right into an unpaved road in front of the school building and continue for about 1.5 km to the top of the hill towards a telecommunication tower. This last part is only accessible with a four-wheel drive car and caution should be taken, especially after rainfalls when deep gullies may have formed on the track.

The viewpoint is located 170 m above the sea where steep rocky cliffs formed. The views over the stark pristine landscape are well worth the effort to get there. Three rock types are characteristic for this area. The oldest ones are grey Permian dolostones of the Saiq formation, overlain by Triassic limestones and dolostones of the Mahil formation. Both formations are metamorphosed. Discordantly overlying Eocene marly

Fig. 41: View into drowned valleys from the Bandar al Khayran viewpoint.

limestones of the Jafnayn formation make up the hills to the north of the viewpoint. The Permian-Triassic boundary is halfway up the mountain (UTM 40 Q 678095 2600635 / N 23°30'21" E 58°44'39"). The rocks at the viewpoint are mainly marbles and calc-silicates belonging to the Saiq formation. White feldspar and orange-brown coloured baryte precipitates occur with large crystals broken along the cleavage planes in cm-wide fractures.

The scenery is dominated by lagoons fringed by mangroves, intertidal mudflats and saltmarshes. Small pocket beaches are located on the outer coast. Most of them are hardly accessible from the landside and are only accessible by boat. This coastal area offers sheltered habitats for corals and the place is well-known for scuba diving and snorkeling.

The coastal configuration with lagoons and tidal creeks (Fig. 41) formed as a result of coastal subsidence as the wadi mouths drowned in the sea (see Hoffmann et al. 2013c).

EP 05

As Sifah
Note: A 4-wheel drive car as well as sturdy shoes are recommended for this route. It includes some difficult parts of scrambling over rocks.
Topics: Eclogite, mica schists, high metamorphic rocks
Locations: a) UTM 40 Q 681796 2594840 / N 23°27'13" E 58°46'47"
　　　　　　b) UTM 40 Q 681504 2592905 / N 23°26'09" E 58°46'35"
　　　　　　c) UTM 40 Q 680745 2593001 / N 23°26'13" E 58°46'09"

Rating: ☺☺

Approach: The road to As Sifah runs through Yiti and Bandar al Khayran (see location map of EP 03, Fig. 39). The relatively narrow road from Yiti to As Sifah passes through a number of beautiful creeks, known as khyran or khor in Arabic. Turn to the left in As Sifah 14 km after Bandar al Khayran. Then drive north along the initially paved then sandy track for 2 km. Attention: Cars may get stuck along the track. At the end of the beach one has to climb about 500 m along the cliffs to the next beach to the north. The path is quite difficult, especially during high tide, because of large boulders and cliff collapse deposits. The eclogite outcrop (location 5a, Fig. 42) is along the coastal cliff of this isolated beach which is around 500 meters long. Location 5b (fig. 42) can be reached from the end of the paved road following a foot path for about 400 meters to the west. Location 5c (Fig. 42) is near the main road.

The village of As Sifah lies on the eastern side of the Saih Hatat bowl along the coastline. The village is well-known for having one of the most beautiful sandy beaches in Oman. This beach and the contrasting distinctive jagged hills attract many visitors particularly during winter. The location is also famous for some eclogite outcrops that are

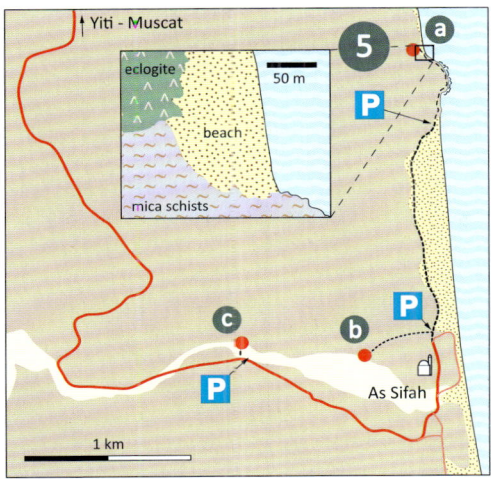

Fig. 42: Location map.

located north of the village at an isolated beach (location 5a; Fig. 42). The eclogite represents a unique opportunity to study high pressure metamorphism in Oman (El-Shazly & Coleman 1990, Searle et al. 1994). The As Sifah area is the only region in Oman where eclogite occurs. The pre-Permian units of As Sifah, which consist of meta-greywackes, meta-volcanics, meta-dolomites and quartzites, are exposed in the core of the dome in central and western Saih Hatat. They are flanked from all sides by metamorphosed mid-Permian to Cenomanian shelf units that almost always dip away from the core of the dome (see Fig. 46, EP 06). Eclogites are derived from gabbroic or basaltic rocks under particular high pressure and low temperature metamorphic conditions which indicate a burial of at least a 50-km depth. The age of the gabbroic or basaltic rocks is radiometrically determined as 110 million years. A minimum age is given for the high-pressure metamorphism which took place in the Late Cretaceous, about 80 million years ago, contemporaneous with the emplacement of the Samail Ophiolite (El-Shazly et al. 2001). Later exhumation in combination with retrograde greenschist metamorphism brought them back to the surface (see chapter 5.3).

The outcrop consists of interlayered metapelites, marbles with phengite, quartz mica schists in all locations and metabasites (location 5a – eclogite; locations 5b, c – glaucophane schist, epidote hornblende schist; Fig. 42) of the Saiq formation. The highest pressure rocks, around 15–20 kb eclogites occur at the deepest structural level in the As Sifah region. The eclogites consist of coarse-grained garnet of 0.5 to 3 cm in diameter, pyroxene, kyanite and lawsonite (Fig. 43). Coarse-grained eclogite facies metabasites show a prograde mineral assemblage of garnet, clinopyroxene (chloromelanite), lawsonite, phengite, glaucophane and rutile. Eclogite facies metapelites show: Quartz,

Fig. 43: Eclogite at the beach of As Sifah (location 5a, Fig. 42).

phengite, garnet, chloritoid, clinopyroxene (aegerine-jadeite) and lawsonite. Pseudomorphs of clinozoisite and white mica after lawsonite enclosed in garnets show a prograde growth zoning (El-Shazly & Liou 1991).

Boulders of iron ore (haematite and magnetite) can be found as surf conglomerate along the rocky part of the coast. They most probably derive from dolerite dikes and sills which intrude the Saiq formation. Furthermore, the metamorphic sediments display very nicely developed mylonitic shear structures such as S-C fabrics, rotated boudins and rotated s- and d-clasts (Fig. 44).

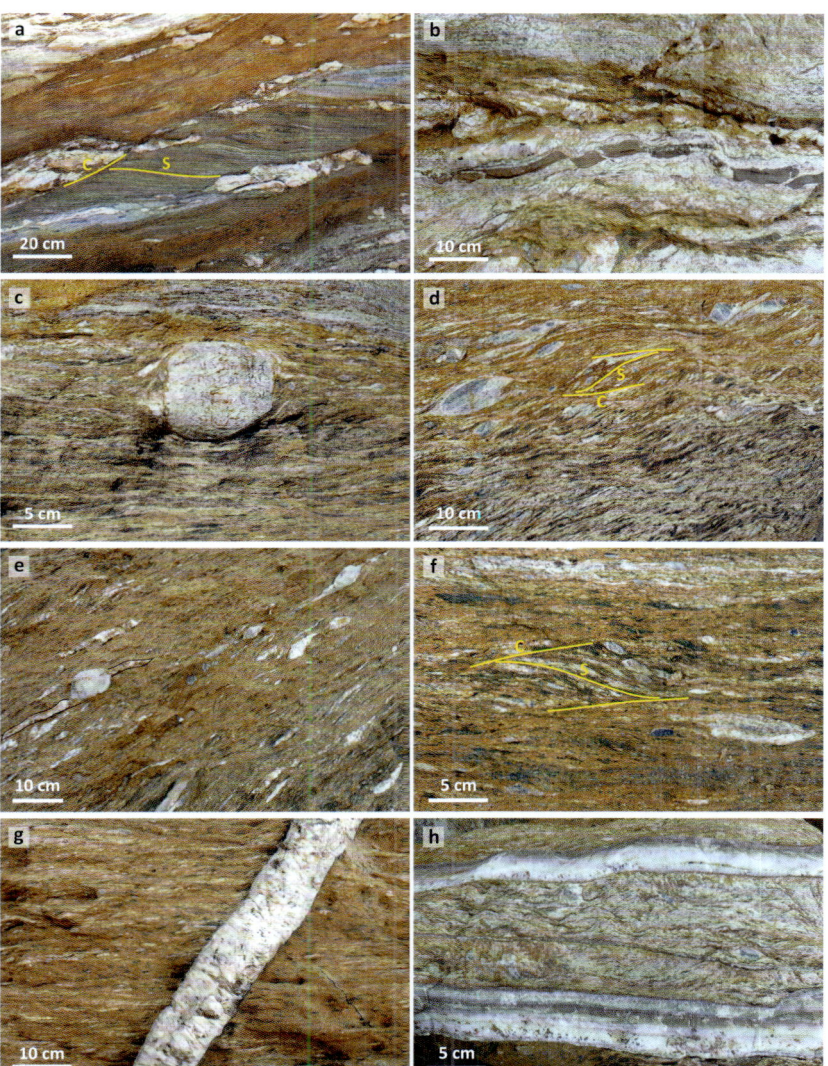

Fig. 44: Various fabrics in metamorphic sediments of the Saiq formation along the cliff at location 5a: a) S-C fabric, b) antithetically rotated asymmetric boudins, c) rotated σ-clast, d) rotated σ-clasts and S-C fabric, e) rotated δ- and σ-clasts, f) S-C fabric, g) quartz filled vein in mica schist, h) quartz ribbons and S-C fabric.

EP 06

Wadi al Mayh
Note: The sheath fold viewed during this trip is described as one of the largest and best exposed in the world.
Topic: Saih-Hatat window, mega-sheath folds
Locations: a: UTM 40 Q 656905 2591896 / N 23°30'18" E 58°37'29"
b: UTM 40 Q 665647 2599125 / N 23°29'37" E 58°37'19"
c: UTM 40 Q 657187 2591860 / N 23°25'44" E 58°32'18"
d: UTM 40 Q 656905 2591896 / N 23°25'45" E 58°32'08"
e: UTM 40 Q 655703 2589615 / N 23°24'32" E 58°31'25"

Rating: ☺☺☺

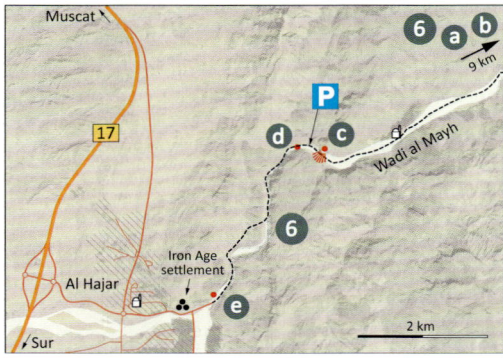

Fig. 45: Location map. For locations of EP 06a and b see EP 03, Fig. 39.

Approach: Follow the main road #17 from Muscat to Sur. Take the first exit to Al Hajar and turn left into Wadi al Mayh. The best outcrops (locations 6c, d) are located directly after leaving the narrow part of the wadi, about 3.5 km after Al Hajar. Locations 6a and b may be reached by continuing on the dirt road into Wadi al Mayh or leaving the road from Hamriyah to Yiti (see EP 03 / Fig. 39) into Wadi al Mayh. After about 1.3 km, you will arrive at location 6a, and after another 3 km, there is a view point (location 6b, Fig. 39) to a large scale fold.

The folds exposed in Wadi al Mayh are part of the Saih Hatat fold-nappe which is a regional high pressure belt in the shelf carbonate units that forms a banana-shaped NE-facing and closing major anticlinal recumbent fold (Fig. 46). They are the structurally highest units within the Saih Hatat window (Miller et al. 1998). Here, the basement and Permian cover rocks are folded around a 15 km long NNE-aligned isoclinal sheath fold located above a major detachment called "upper plate – lower

Field sites

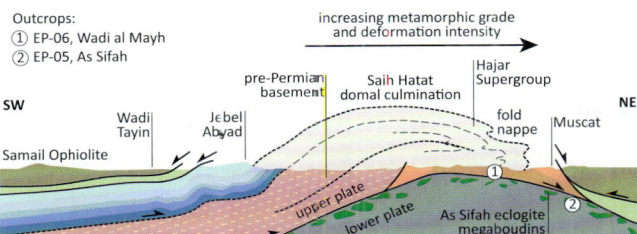

Fig. 46: Cross section through the Wadi al Mayh mega-sheath fold with the location of the outcrops in Wadi al Mayh (1) and the As Sifah eclogite (2) (after Searle & Alsop 2007).

plate discontinuity" (Gray et al. 2005, Miller et al. 2002). The two plates display differences in metamorphic grade and structural style. Whereas the rocks below this discontinuity are highly metamorphic (compare EP 05, As Sifah), the rocks above show ductile deformation by internal folding, duplexing and imbrication (Searle 2007). The structures are the result of the exhumation from a NE-dipping subduction zone which was accompanied with high ductile strain with major shear zones. The subduction of the leading edge of the Arabian Plate beneath the oceanic hanging wall is supposed to have happened during the Campanian, around 15 million years after the formation of the ophiolite.

One of the highlights in the exposures along the Wadi al Mayh gorge is a unique mega-sheath fold in Permian, and possibly Triassic, shelf carbonates that consists of tight to isoclinal folds bounded by low-angle detachments, which have been folded

Fig. 47: Mega-sheath fold in Permian and Triassic limestones and dolomites in Wadi al Mayh (location EP 06c, Fig. 45).

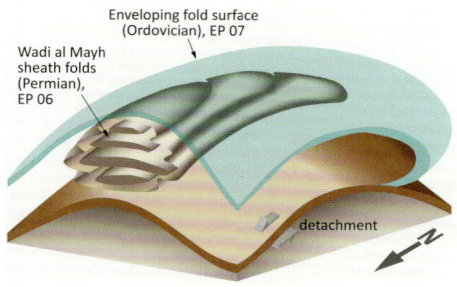

Fig. 48: 3D block diagram model of the geometry of the Wadi al Mayh mega-sheath fold (modified after Searle 2007).

within the mega-sheath fold (location EP 06c; Figs. 45, 47, 48). This sheath fold is described as one of the largest and best-exposed in the world.

There are other large-scale and medium-scale folds within the wadi (locations EP 06a, b [see EP 03, Fig. 39], Fig. 49; location EP 06e, Fig. 45), all interpreted as late-stage back folds that formed after peak metamorphism and during the cooling of the upper plate at the late stages of the obduction. A large parallel fold in thick bedded limestone is well exposed on the opposite side of location EP 06c (Fig. 45). To the west, more incompetent sedimentary layers of siltstone and marl, with a more complex outline of folding, can be found.

The narrow valley of Wadi al Mayh which follows towards the southwest (starting at location 6d) is dominated by more competent layers of limestone. Here the metamorphosed interbedded limestone and dolomite show impressive examples of boudinage structures. These formed as a consequence of stretching deformation (e.g., well exposed at location EP 06d; Figs. 45, 50). The most competent layers of dolomite form the

Fig. 49: Large scale fold in Wadi al Mayh (location EP 06b, Fig. 39).

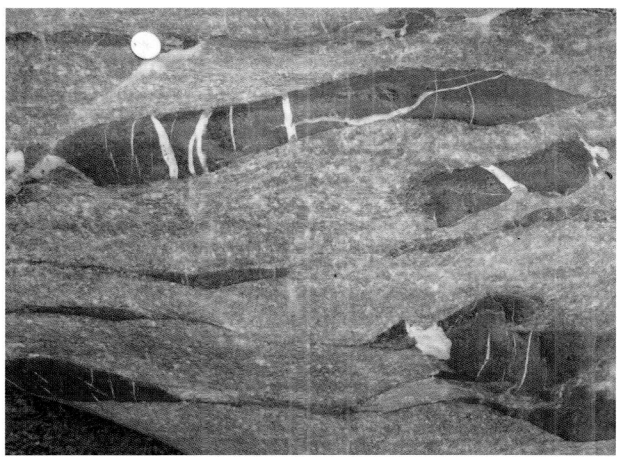

Fig. 50: Boudinage of metamorphosed interbedded limestones and dolomites in Wadi al Mayh (location EP 06d, Fig. 45).

boudins in a matrix of carbonate, and fractures in the dolomite are filled with calcite precipitations.

Remnants of a Late Iron Age settlement can be observed on a hillside (UTM 40 Q 655112 2589389 / N 23°24'25" E 58°31'04") in the neighborhood of the modern village of Al Hajar where the narrow wadi opens (Fig. 45). Oman's Samad complex, the Late Iron Age of Oman (150 BCE – 300? CE), is still only poorly studied and the site has not been scientifically documented so far (Yule & Kervran 1993).

EP 07

Wadi Amdeh
Topic: Ordovician sediments of the Amdeh Formation, ripple marks
Location: UTM 40 Q 641231 2579512 / N 23°19'08" E 58°22'52"

Rating: ☺

Approach: Follow the main road #17 from Muscat to Sur. Take the exit at Al Hajar (UTM 40 Q 641280 2579694 / N 23°19'14" E 58°22'54"). Turn right and follow the paved road towards the southwest. After about 6 km, turn right onto an unpaved road through a series of Late Proterozoic schists and follow this road for about 5 km. Turn left (at UTM 40 Q 641280 2579694 / N 23°19'14" E 58°22'54") on the soft shoulder of the road entering a narrow dirt road (at UTM 40 Q 660281 2582217 / N 23°20'30"

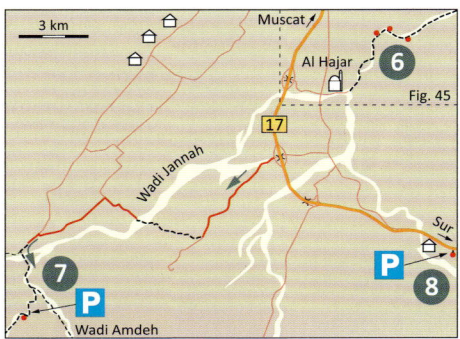

Fig. 51: Location map

E 58°34'03"). Turn left onto a straight road for 3.5 km, then turn left again. The tarred road changes into a dirt road after 700 m and continues for about 1.2 km. Turn left and follow the dirt road into the Wadi Amdeh for about 1.5 km, then turn right into a small wadi on the right-hand side. The outcrop is about 1.8 km after the entrance into the small wadi.

Fig. 52: Ripple marks in sandstones of the Amdeh formation, Ordovician.

The rocks in this outcrop belong to the Early/Middle Ordovician Amdeh formation. This mainly clastic sequence of shale and quartzite is weakly metamorphosed. The so-called Lower Quarzite crops out on the southern side of the wadi. The rock is evenly bedded and the layers dip around 25° to the North. The strata are inclined in the same way as the slope (dip slope) and layers of decimetre thick quartzite are breaking off. Bioturbation in the form of *Skolithos-Cruziana* ichnofacies is described by Villey et al. (1986b). The rock surfaces, and as negative imprints also the base of the layers, are characterised by ripple structures (Fig. 52). The individual ripple marks are 1–2 cm high and the crests spaced 3 to 5 cm apart. The ripples are symmetric, indicating formation by waves rather than by currents. The ichnofossils are from both, suspension- and deposit-feeding organisms. Hence ripple marks and ichnofossils indicate deposition in a shallow epi-continental sea. The sandy nature of the deposits points to nearshore, shore-face or barrier environments. Centimetre-sized nodules are observable within the ripple troughs (Fig. 52). Their origin as faecal pellets remains speculative.

EP 08

Moonscape near Al Hajar
Topic: Late Proterozoic phyllites of the Hatat formation
Location: UTM 40 Q 660080 2582067 / N 23°20'25"E 58°33'56"

Rating: ☺

Location map see EP 07, Fig. 51.

Approach: Follow the main road #17 from Muscat to Sur. About 9 km after the exit to Al Hajar, turn right on the soft shoulder of the road and enter a narrow dirt road (at UTM 40 Q 660281 2582217 / N 23°20'30" E 58°34'03"). Follow the dirt road for 100 to 200 m into the outcrop.

The small hills at this outcrop consist of greenish banded schists or phyllites, which are mainly derived from a Late Proterozoic greywacke (Le Métour & Villey 1986b). They belong to the Amarat Member of the Hatat Formation. The Hatat Formation is the oldest stratigraphic unit exposed in the Saih Hatat area that represents the autochthonous continental basement of Oman. The greywacke was originally formed as silicic tuffaceous deposit in a volcanic environment, probably at an active continental margin. Although the schistosity and metamorphic overprint have blurred most of the sedimentary structures, bedding is preserved in dm- to m-scale. Hydrothermal mineralisations of cm to dm thick white and brownish bands consist of mobilised quartz (SiO_2) and siderite ($FeCO_3$), and cut through the phyllitic rocks in the outcrop. Rock crystals, a clear and colourless variety of quartz that is up to several centimetres in length, may be found on the surface, weathered out of the hydrothermal veins.

Fig. 53: Kink bands in phyllites of the Amarat Member (Hatat formation), Late Proterozoic.

The phyllites are characterised by intense schistosity. Two nearly vertical foliation planes result in a penetrative pencil cleavage structure. Kink bands and kink folds of cm-size are typical for these rocks (Fig. 53). They represent a late stage of deformation which developed at the transition from brittle to ductile deformation.

EP 09

Wadi Daiqa
Topics: Palaeozoic fossils, Mesozoic carbonates, faults, scenic view over the dam – especially after rain
Location: UTM 40 Q 689601 2555227 / N 23°05'41" E 58°51'04"

Rating: ☺☺

Approach: From Muscat, follow the express way #17 to Sur for about 85 km. Take the exit just 1 km before the village of Ramlah (at UTM 40 Q 692528 2567076 / N 23°12'03" E 58°52'50"). Follow the sign posts to the Wadi Daiqa Dam and turn left at a roundabout. Follow the road for 1.9 km and then turn left. There is another left turn after 3 km. Drive on for 13 km and take a left turn at UTM 40 Q 689601 2555227 / N 23°05'41" E 58°51'04" just before entering the village of Al Ghubaira. Follow the road for about 1.5 km to get to the main overview of the dam. A parking area is located uphill close to the dam control centre.

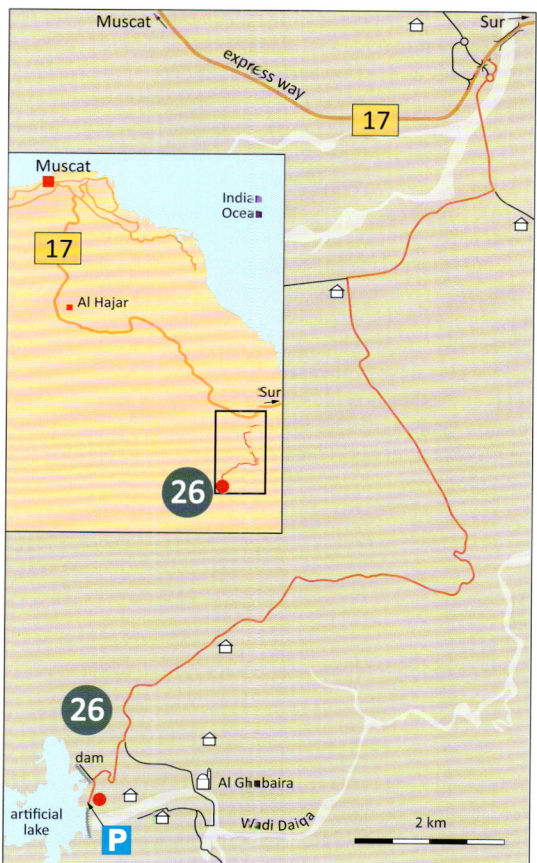

Fig. 54: Location map

The big artificial lake upstream of Wadi Daiqa is surrounded by siliciclastic Palaeozoic and carbonatic Mesozoic rocks that are deformed by a number of major fold and fault structures (Heward 2012). The exposures of the Amdeh formation, which is approximately 3400 m thick, represent the most prominent outcrops in the area. The formation consists dominantly of shallow marine sediments (cf. EP 07) and dates back to the Ordovician period (Early Palaeozoic). The formation is divided into five members, from Amdeh 1 at the base to Amdeh 5 at the top. Although the nature of the material from which the dates have been derived lacks high resolution, the earliest Amdeh specimens are at least contemporaneous with the oldest occurrence of *Sacabambaspis* which is an extinct genus of jawless fish known also from Australia and South America. The sediments of the Amdeh formation represent shallow water deposits with a mixed *Skolithos*, common trace fossils that consist of vertical cylinders and *Cruziana*, trace fossils of trilobites ichnofacies containing trace fossils of both suspension- and deposit-feeding tracemakers (e.g. *Cruziana* and *Daedalus*). In association with trilobites (*Neseuretus* and *Ogyginus*), the bivalved mollusc *Redonia*, crinoid remains and orthoconic nautiloids, are indicative of nearshore conditions.

The fragments of "*Sacabambaspis*" in Wadi Daiqa are the oldest known group of vertebrates with extensive biomineralisation of the dermo-skeleton. Their presence in Oman greatly extends the palaeogeographical distribution of the clade around the Gondwanan margin. Clade is a taxonomic term that refers to a group of organisms sharing a common ancestor. A thick succession of Al Khlata glacio-lacustrine formation, including diamictites, also crops out in the Wadi Daiqa inlier (Heward & Penney 2014). This formation is a well-known oil reservoir in central and southern Oman and was deposited when Oman was located close to the South Pole in the Carboniferous period more than 300 million years ago. The grey overlying carbonate rocks are Late Palaeozoic (Permian) or Mesozoic in age. They are rich in shallow-marine fossils, including corals, crinoids and molluscs.

EP 10

Bimmah Sinkhole
Topic: Karstification
Location: UTM 40 Q 712322 2548990 / N 23°02'09" E 59°04'19"

Rating: ☺☺

Approach: Leave the main road #17 from Muscat to Sur at the exit Bimmah Sinkhole/ Hawiyat Najm Parc. Follow the road towards the coast and turn left twice until you reach the entrance to the Hawiyat Najm Parc. A short path leads to the sinkhole. The Arabic name "Hawiyat Najm" refers to a local legend and translates to "meteorite impact crater". However, the sinkhole represents a collapsed cave in a karstified area and formed due to chemical dissolution of carbonate rocks. Karstification describes the process of limestone dissolution by meteoric water.

The sinkhole, or doline, is situated within an elevated abrasion platform of Eocene limestone (Figs. 56, 57). It has a slightly oval shape with a diameter of 65 m at the widest point and 50 m at the narrowest. It has a depth of about 25 m. The water in the sinkhole is a mixture of sea water and groundwater. The hydraulic contact between salty sea water and groundwater contributes to the erosional effect and plays an important role in karstification in this area. The Eocene limestone is overlain by a 5 m thick layer of terrestrial gravel resulting from the abrasion of the limestone platform. This indicates sea-level changes caused at this location mainly by tectonic

Field sites

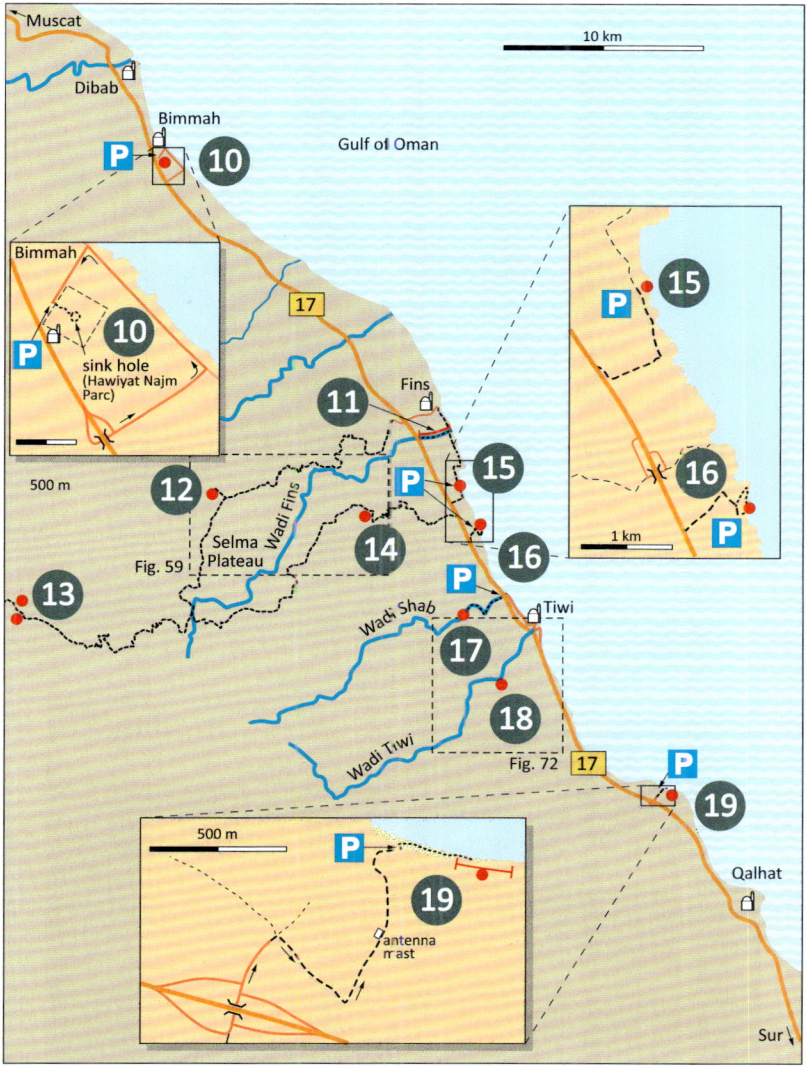

Fig. 55: Location map

Fig. 56: The Bimmah Sinkhole within Eocene limestone.

Fig. 57: The Bimmah Sinkhole, view from below.

uplift. There are steps leading down into the sinkhole where it is possible to swim in the water.

EP 11

Fins unconformity, Wadi Fins
Topic: Unconformity Eocene/Pleistocene
Note: Flintstone flakes represent the remnants of a Neolithic flint workshop
Location:
Cross section from UTM 40 Q 727400 2535610 / N 22°54'47" E 59°13'02"
          to   UTM 40 Q 725800 2535231 / N 22°54'35" E 59°12'06"

Rating: ☺

Location map see EP 10, Fig. 55.

Approach: Leave the main road #17 from Muscat to Sur at the exit to Fins and follow the road to Fins. At the end of the road turn right before entering the small village. Follow the road along the coast for 1.2 km. The profile starts at the mouth of Wadi Fins and continues 1.7 km upstream up to the bridge on the highway #17.

Exposures along the wadi reveal an angular unconformity between the lower Eocene limestone, that dips by 25° towards the sea, and the horizontally bedded Quaternary cover (Fig. 58). The latter compromises of beachrock deposits and limestone made up of coralline algae. The cover sediments become dominated by fluvial gravel further towards the highway. In-situ remains of the gastropod *Pleuroploca trapezium* can be

Fig. 58: Angular unconformity of Pleistocene sediments discordantly overlying tilted Eocene strata.

EP 12

Majlis al Jin cave, Selma Plateau
Topic: Karstification
Note: A four-wheel drive car is essential due to the very steep and unpaved road.
Location: UTM 40 Q 716016 2531959 / N 22°52'53" E 59°06'20"

Rating: ☺☺

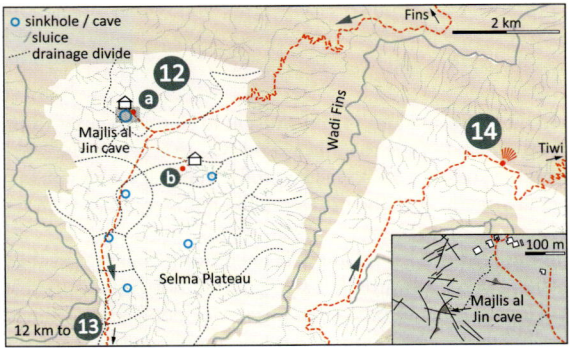

Fig. 59: Location map. For location see also the map at EP 10, Fig. 55. The inset shows intersections of fractures around the entrances into the large chamber of the cave.

Approach: Leave the main road #17 from Muscat to Sur at the exit to Fins and turn right at the sign to Wusal, following the road uphill for about 17 to 18 km. Only the first approximately 700 m is asphalt. The unpaved road is extremely steep in parts, rising up to an angle of more than 50 % (i.e. more than 25°)! Therefore, a four-wheel drive car is essential. Turn right at the fork. In 600 m you will arrive at the Majlis al Jin cave. Majlis al Jin is a cave system with an entrance at 1390 m above the sea level, which formed on the limestone plateau known as Selma Plateau or Bani Jabir Plateau. The mountainous region of the plateau called Jebel Khadar is broad and elevated up to 2223 m above sea level at the highest point. The plateau is mainly made up of fossiliferous carbonate rocks of Middle Eocene age which were formed as shelf deposits of the Hadhramaut group, Dammam formation.

Nummulites are especially abundant in the rocks outcropping at the location of the cave's entrance. Davison (1990) argues that the cave's ceiling is made up of marl and

Fig. 60: Sinkhole on top of the Selma Plateau (location 12b, Fig. 59).

that the additional clay content makes the rocks more resistant to weathering. If that had not been the case, the cave would have collapsed. Sinkholes are common in the deeply dissected plateau area, indicating intensive karstification (Fig. 60). The rocks are folded, faulted and fractured and the three entrances to the chamber formed where the fractures intersect (inset in Fig. 59).

The Arabic name "Majlis al Jin" translates to "the meeting room of the spirits". It was named by the geologists Cheryl Jones and Don Davison, who first explored the cave system in 1983 together with Doug Green. No official name existed at that time. However, local tribesmen call the place Khoshilat Maqandeli. The dimensions of the single chamber cave are enormous and are measured as 310×225 m. The floor area is calculated as 58000 m^2 and the volume as 4×10^5 m^3. The domed ceiling is up to 140 m above

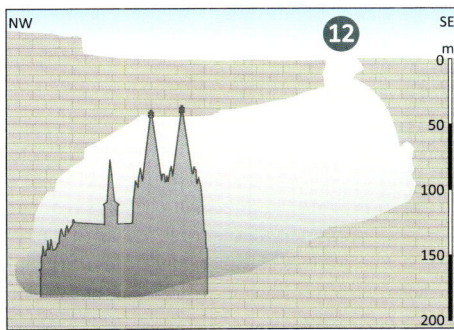

Fig. 61: Sketch to illustrate the size of the Majlis al Jin cave with the contours of the Cologne Cathedral (location 12a, Fig. 59).

Fig. 62: Aerial view of the cave's entrance (location 12a, Fig. 59). Please note the car for scale.

the cave's floor (Fig. 61). The deepest part of cave is measured as 178 m below the entrance. The Cologne Cathedral (Germany) that has a height of 152 m was added for scale to Fig. 61 to illustrate the chamber's size.

The modern climate of the plateau is characterised by sporadic precipitation events with a total annual amount of 180 mm. The surface water is immediately transferred into the karst system following precipitation events. A perennial lake develops in the cave, leaving behind a mud flat. Sediment input into the cave system is furthered by dust brought in by wind as well as debris that fall through the holes. Huge talus heaps of debris containing bones and other remains of organisms are described below the entrance holes (Davison 1990). There are only a few stalactites documented.

The cave system must have formed when the climate was substantially different in the past. Climate oscillations between dry and wet periods are characteristic for the Arabian Peninsula and are linked to global climate changes. Periods of higher latitude glaciation are known to correlate with arid conditions and the humid periods primarily coincided with interglacial periods. The cave system is the result of numerous wet periods; the most recent one culminated approximately 6000 years ago.

Access to the cave is only possible by rappelling down 120 m from the ceiling (Fig. 62) and is restricted to permit holders by the Ministry of Tourism. Austrian born Felix Baumgartner dared a base jump into the cave. This method of entering is not recommended!

EP 13

Tower tombs, Selma Plateau
Note: Due to the steep gradient of the road, a four-wheel drive car is essential.
Topic: Historical site, rare archeological remains of well-preserved tower tombs
Location: UTM 40 Q 710915 2524543 / N 22°48'55" E 59°03'18"

Rating: ☺☺

For location map see EP 10, Fig. 55.

Approach: Leave the main road #17 from Muscat to Sur at the exit to Fins and turn right, following the road uphill for about 17 to 18 km. Only the first approximately 700 m is asphalt. The unpaved road is extremely steep in parts, increasing to an angle of more than 50 % (i.e. more than 25°)! Therefore, a four-wheel drive car is essential. Turn left at the fork and follow the road another 16 km. The tower tombs are along the road.

The Selma Plateau, otherwise known as the Bani Jabir Plateau, is broad and formed of Eocene limestone. The rock formations are intensively karstified (see EP 12) and, therefore, there are almost no sources of fresh water, although the annual precipitation is comparatively high at 180 mm. Soils are also not present and hence agriculture is restricted to goat keeping. The plateau is remote and almost uninhabited nowadays. Maybe surprisingly though, it is very rich in spectacular archaeological remains. Known only to local people, the large number of tower tombs were accidentally discovered in the early 1990s and subsequently scientifically described, mapped and catalogued in 1995 by Paul Yule and Gerd Weisgerber. The most important artefacts are assembled on a crest around 1700 m above sea level that is known as known as the "Necropolis at Shir" (Yule & Weisgerber 1998).

There are remains of more than 50 tower tombs in various stages of preservation. All of the structures are constructed out of the local Tertiary limestone, partly unworked and partly finely dressed. The structures also differ in type and size. Yule & Weisgerber (1998) differentiated 5 types. The largest towers are more than 7 m high. The majority of the towers is located at the edges of bluffs and serve as landmarks which are visible from the distance. Both single and double wall constructions are noticeable. The towers have east-facing entrances at ground level and are circular in shape. All structures had been plundered by the time of the first scientific investigations; no skeletons were preserved.

Archaeological evidence suggests that the structures were erected as memorial buildings for one or more deceased persons. The tombs are arranged in four groups, each of those dominated by one tomb that is markedly larger, better preserved and of higher quality in general. One of these is Shi1 (this location), located at 1758 m above sea level, with a base of 6 m diameter and a preserved height of 5.47 m (Figs. 63, 64). The

Fig. 63: Tower tombs at the western margin of the Selma Plateau.

Fig. 64: Aerial view of the tower tombs with Shi 1 in the foreground.

tower rests directly on the bedrock. The base of this structure is built of stones sized on average 0.50 × 0.20 m, overlain by stones measuring c. 0.35 × 0.12 m. The outer side of the stones is convex and flaking marks are often visible, indicating that the building material was carefully worked on by skilled craftsmen. Entrances usually face east.

None of the material discovered during the mapping project by Yule & Weisgerber (1998) could be used for absolute dating of the structures. Age constraints are obtained by comparisons with other structures. Some of the towers appear to be closely related to beehive tombs of the Hafit period from the first half of 3rd millennium BCE, while other have more similarities to buildings categorised as the Umm an-Nar period which was the second half of 3rd millennium BCE. Whatever the case, it is apparent that these impressive structures roughly date back to the times when the first pyramids where built in Egypt. Further investigations will have to prove whether or not the tombs were erected by a local population or by people of foreign origins. Myths explain the building of the towers (Yule & Bergoffen 1999).

EP 14

Viewpoint on Pleistocene terraces from the Selma Plateau
Note: A 4W-drive vehicle is essential for this trip due to the steep gradient of the unpaved road.
Topic: Pleistocene terraces along the coastline of the Gulf of Oman
Location: UTM 40 Q 722975 2531076 / N 22°52'22" E 59°10'23"

Rating: ☺

For location maps see EP 10, Fig. 55, and EP 12, Fig. 59.

Approach: Leave the main road #17 from Muscat to Sur 5.5 km after the exit to Fins, and turn right following the unpaved road uphill. The unpaved road is extremely steep in parts, rising up to more than 50 % (i.e. more than 25°)! Therefore, a 4-wheel

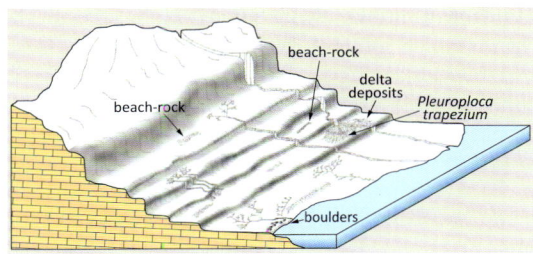

Fig. 65: Structural overview of the uplifted stair-cased terraces close to Fins, view NNW.

Fig. 66: Panoramic view of the different terrace levels (view from Fins towards west).

drive vehicle is essential. The viewpoint is about 9 km along the road on top of the Selma Plateau. Another possibility is to continue from EP 12, turning left at a fork (at UTM 40 Q 715326 2526074 / N 22°49'42" E 59°05'53") in a small village. The viewpoint can be reached after about 13 km, following the unpaved road towards the north-east.

The coastline shows geomorphologic evidence of crustal uplift. Stair-cased coastal terraces parallel to the coastline (Figs. 65 and 66) are encountered in elevations up to 200 m bordering the Selma Plateau with a height of 1500 m, located in the west. The passive margin of the Arabian Peninsula has relatively steep continental slopes with an average dip exceeding 5.5°. The upper most terraces are erosional, whereas the lower ones are depositional in style. The erosional terraces are cut into Paleocene to Early Eocene limestone formations. The limestone is karstified; sink holes (see EP 10, Fig. 55) and caves (see EP 12, Fig. 59) are common features. Mollusc and coral remains as well as beach-rock are encountered on the terrace surfaces. The formations are dissected by NW-SE trending faults. Some of the terraces are very pronounced features in the landscape and easy to trace, others are partly eroded and preserved as remnants only. Wadis cut the terraces. Some of the wadis drain the adjacent Selma Plateau and some developed on the different terrace levels. The dating of the different terraces is critical for a quantitative approach. In general the terraces become younger as the elevation decreases. Currently, dating results (Hannss & Kürschner 1998, Kellerhals 1998) are inconclusive and do not allow a palaeogeographical reconstruction or further interpretation.

EP 15

Tsunami boulder train
Topic: Tsunami history
Location: UTM 40 Q 728093 2532908 / N 22°53'18" E 59°13'24"

Rating: ☺☺

For location map see EP 10, Fig 55.

Approach: Leave the main road #17 from Muscat to Sur 4.5 km after the exit to Fins (at UTM 40 Q 727496 2532156 / N 22°52'54" E 59°13'03") and turn left, following an unpaved road to the coast. Turn left after 800 m and follow the road parallel to the coast for about 700 m to the outcrop.

The rocky coast along the strip between Sur and Quriyat appears relatively "cleaned", i.e. only sparse rubble and plant remains are found approx. 10 m to the cliff front. Huge, angular blocks of Tertiary limestones and Quaternary beach rocks are found in this area (Fig. 67). Most of these blocks are tilted, toppled and partly over-turned, which is evidenced by bio-erosional features on the lower side and smooth surfaces (rock pools). Also, impact marks can be observed on the surface of the bedding planes and on the boulders. The origin and provenance of the boulders suggest low transportation distance. They derive from the cliff top of Tertiary limestone and Quaternary beach rock nearby. These rocks show joints, fractures and bedding planes that pre-shape the boulders.

The thick banked (0.5–1.5 m) angular boulders are made up of Eocene limestone. These are partly karstified and decorated with marine sessile fossil remains. Whereas sub-angular, platy or elongated boulders with a thickness of approximately 0.5 m consist of Quaternary beach rock, which covers the Eocene limestone close to the cliff. Oysters are attached to rounded blocks indicating an origin from the immediate shoreline, where these organisms usually live. This is also evidenced by *Lithophaga sp.* borings, which also point to a tidal position before transport.

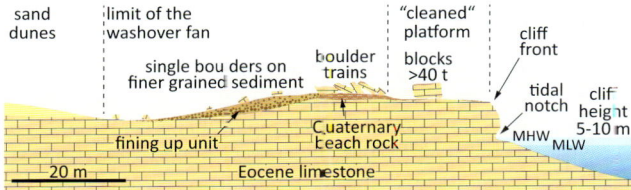

Fig. 67: Schematic cross section of the platform and washover deposits with boulder trains and mega-boulders. MHW = mean high water, MLW = mean low water (tidal range).

Most of the boulders are aligned with the a-axis towards N30E. They are imbricated and form so-called boulder trains. The N30E direction is at a high angle to the present-day coastline stretching approximately N-S. The largest boulders were also measured to obtain weight information. The largest boulder found so far has a mass of 120 tons (see EP 16).

The process that led to the displacement of the boulders must have been a high energy wave event and could have been either a storm or a tsunami (Hoffmann et al. 2013a, Koster et al. 2014). However, boulder movements were not observed during the recent tropical cyclone impacts in 2007 and 2010. It is therefore concluded that these are tsunamigenic boulders. The dating of the respective events is part of ongoing research.

EP 16

Tsunami boulders
Note: It is possible to do this trip in a normal sedan car, with some walking. However, a four-wheel drive car is recommended.
Topic: Tsunami deposits
Location: UTM 40 Q 729217 2530530 / N 22°52'01" E 59°14'03"

Rating: ☺☺☺

For location map see EP 10, Fig. 55.

Approach: Leave the main road #17 from Muscat to Sur at the exit to Wadi Shab and then return onto the highway towards Muscat. Pass the Wadi Shab Hotel and leave the highway after 2.1 km, turning left onto a dirt road towards the coast. This turn is just before the barrier by the side of the highway starts. Either start walking from here or continue by four-wheel drive car. Turn right after a steep part, terrace scarp, and go to the small gravel pit at UTM 40 Q 729128 2530501 / N 22°52'00" E 59°13'59".

The car park is located within a small gravel pit where material for construction purposes was mined. The remaining material on the eastern or seaward side of the pit allows an insight into the stratigraphy of shore-parallel coastal ridges (compare with EP 15). The ridges consist of reworked beach material, mainly sand with floating boulders and gravel with bivalves like oysters and *Tridacna sp.*, scaphopods, gastropods and coral fragments (Fig. 68). The marine faunal remains testify to a near-shore origin. Also *Lithophaga sp.* borings in limestone pebbles support this assumption. The entire deposit apparently has an inverse grading, as the largest boulders are found on the top of it. However, the sedimentary sequence shows internal fining-up cycles. One boulder (c. $1.0 \times 1.0 \times 0.8$ m) with attached oysters was found 80 m from the cliff edge and 8 m above mean sea-level. This may reflect the maximum transport distance for boulders.

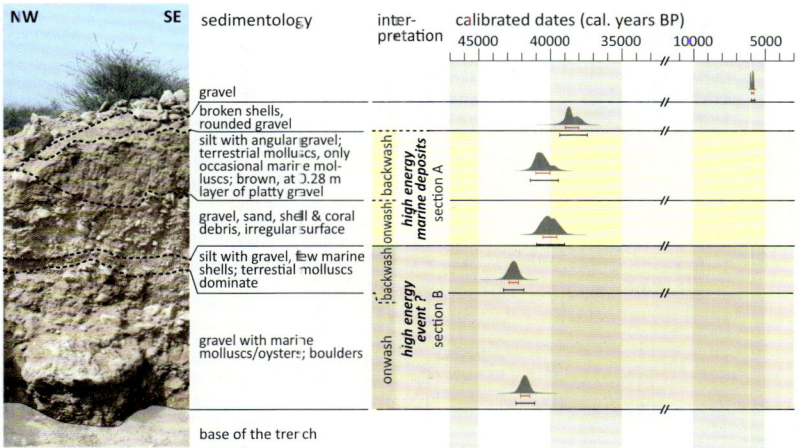

Fig. 68: The trench reveals sedimentological evidence for possibly two high energy tsunami waves with onwash (marine composition) and backwash (mixed marine/terrestrial composition). Ages for each layer are given in calibrated years BP.

A short walk towards the cliff involves a climb over the boulder ridges onto the cliff platform. This platform is made up of Eocene limestone. Remnants of lithified beach deposits in the form of beach rock cover the limestone in places. The platform formed by a combination of wave action and bioerosion (cf. Taboroši & Kázmér 2013). Uplift

Fig. 69: Tsunami boulders resting on an uplifted and clean surf zone. The shift of the boulders behind the surf zone and the clearing of the frontal part of the steep cliff is the result of a tsunami event.

Fig. 70: The largest tsunami block found along the coastline has a mass of 120 tons.

of the crust then resulted in emergence of the platform as a terrace well above present high tide level. The terrace surface is barren (Fig. 69); there are no deposits except for some isolated blocks, several meters in diameter. The angular boulders at (UTM 40 Q 729217 2530530 / N 22°52'01" E 59°14'03") are made up of Eocene limestone and the largest one has dimensions of 7 × 6 × 1.5 m and a mass of 120 tons (Fig. 70). The blocks rest on the lithified beach rock and hence must have been moved against gravity. They were quarried at the cliff edge.

A rounded limestone boulder can be observed 250 m further north (UTM 40 Q 729203 2530788 / N 22°52'09" E 59°14'02"). The rounding as well as the presence of sessile marine organisms, e.g. oysters, bryozoans, *Lithophaga*, indicate an intertidal origin of this boulder.

It is concluded that the fine grained material which makes up the coast parallel ridges as well as the blocks and boulders were deposited during tsunami events. The most likely source for the tsunami is the Makran Subduction Zone which defines the northern limit of the Arabian Plate (see Fig. 16). The last significant earthquake (M 8.1) that resulted in a tsunami occurred on 27 November 1945. However, the wave impact along the Omani coast was comparatively small (Hoffmann et al. 2013b). The deposits along this coastal section are indicative that larger events than 1945 can be expected.

EP 17

Wadi Shab
Note: The path into the wadi is stony and normal flip-flops or delicate sandals are unsuitable footwear.
Topic: Karstification, doline, rock pools suitable for swimming, terraced plantations
Location: UTM 40 Q 730496 2527460 / N 22°50'21" E 59°14'46"

Rating: ☺☺☺

For location map see EP 10, Fig. 55.

Approach: Leave the main road #17 from Muscat to Sur at the exit to Tiwi and follow the road through the village and down into the Wadi Shab. There is a carpark under the highway bridge. To cross the water-bearing wadi, it is recommended to pay a boatman of one of the boats at the entrance to Wadi Shab. Follow the path into the wadi for about 2 km until you reach a pool (at UTM 40 Q 728800 2526908 / N 22°50'03" E 59°13'46"). Here you can take a swim in a series of pools that are situated upstream for about 300 m. There is also a cave which can be accessed at the end of the row of pools.

Fig. 71: Pool in the Eocene limestones of Wadi Shab.

Wadi Shab is one of Oman's major tourist attractions and is mentioned in every guidebook. It certainly warrants a visit as it truly is one of the most beautiful wadis in the country. The impressive deeply cut canyon with its steep rock faces reaching up to 200 m is accessible only on foot (Fig. 71). The 45 min. walk of 2 km leads to several pools and runs mainly on the wadi bed.

The canyon is cut into Eocene limestone. They appear massive-nodular, yellow to brown in colour, and were shaped by the erosional forces of water. Wadi Shab hosts a perennial stream with a changing water level, depending on precipitation. Additionally, the water is responsible for the creation of an open karst system. Due to carbonic acid weathering, caves, overhangs, and massive block falls are observable all along the walk.

The wadi has spectacular hanging gardens, terraces which are farmed well above the wadi's floor. The water used for irrigation flows through falaj systems onto the fields. Most of the terraces including the soil and vegetation were destroyed by the tropical cyclone Gonu in 2007. Some date palms survived and the terraces are now being re-cultivated.

EP 18

Wadi Tiwi
Topic: Karstification, Samail Ophiolite
Location: UTM 40 Q 730458 2523304 / N 22°48'05" E 59°14'42"

Rating: ☺☺

Approach: Leave the main road #17 from Muscat to Sur at the exit to Tiwi and follow the road into the wadi by passing under the massive highway bridge. Park the car after driving along the narrow lane through the village.

Wadi Tiwi is a deeply cut canyon which runs parallel to Wadi Shab only 4 km to the north (compare EP 17). It also developed in a similar way. The stop described here is in the lower part of the wadi. The wadi is famous for its cultivated plantations which are slowly recovering after the severe impact of cyclone Gonu in 2007. The irrigation is managed with falaj systems. The amount of water depends on the season and the timing of the last rainfall events. There may be nice pools which offer refreshing bathing opportunities, especially after rain.

The Eocene sequence exposed is very similar to Wadi Shab. In contrast, here the underlying ultramafic rocks of the Samail Ophiolite are exposed as well. The rocks crop out in the wadi bed (Fig. 73) and the transition to the overlying Eocene limestone is clearly visible along the valley's sides as the different lithologies have a sharp contrast in colour. The ultramafic rocks are serpentinised and appear heavily brecciated (Fig. 74).

Field sites 105

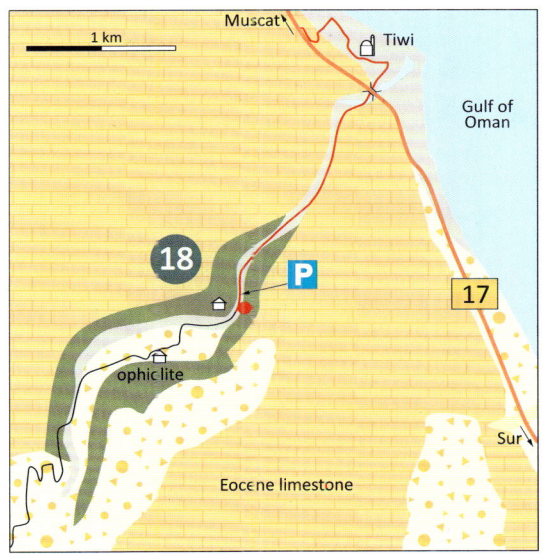

Fig. 72: Location map, see also Fig. 55 for an overview map.

Fig. 73: Ultramafic rocks of the Samail Ophiolite exposed in the river bed of Wadi Tiwi.

Fig. 74: Serpentinised and brecciated ultramafic rocks belonging to the Samail Ophiolite.

Fig. 75: Imbricated concrete slabs: evidence for the impact of tropical cyclone Gonu in 2007.

Note the imbrication structures of concrete slabs that formed by erosion of the road during cyclone Gonu in 2007 (Fig. 75). They were incorporated into the concrete of the new road.

The end of the drivable track is around 6 km further into the wadi. Here the hiking path to Wadi Bani Kahlid starts. If you follow this path for 15 minutes you will reach nice pools which have water all year round.

EP 19

Coral Reef near Tiwi
Topics: Fossil Quaternary coral reef underneath fluviatile conglomerates
Location: UTM 40 Q 739122 2517902 / N 22°45'05 E 59°19'43"

Rating: ☺☺

Location map see EP 10, Fig. 55.

Approach: Follow the highway (#17) from Quriyat to Sur. Take the exit Fayah (at UTM 40Q 738030 2517397 / N 22°44'49" E 59°19'04") and turn towards the coast underneath the highway. The tarmac road ends after 300 m; continue on the graded road towards the coast. Leave the car on top of the cliff (UTM 40 Q 738883 2517925 /

Fig. 76: Fossil Quaternary coral-reef at a cliff south of Tiwi capped by a conglomerate.

N 22°45'06" E 59°19'34") and walk down the path to the gravel beach. The outcrop is located 250 m east.

The coastal area between Quriyat and Sur is currently uplifting as indicated by a flight of raised coastal terraces. The terraces are cut in the Eocene limestone. Whereas the upper terraces are erosional, the lower one is depositional. This outcrop offers the unique possibility to study the recent coastal evolution. We have to assume that the rocks are of Quaternary age. However, they are not dated as yet. The outcrop is located at the mouth of a wadi cutting through the terraces. An intercalation of marine and terrestrial deposits is exposed in a cliff section.

The surface of the terrace was mined for gravel when the highway was constructed from 2005 to 2008. The walk along the gravel beach leads to the base of a coastal cliff which is 9–12 m high. The outcrop shows an impressive succession of coral-reef units and conglomerates. The fossil Quaternary coral-reefs are very well exposed in life position at the wave-cut cliff.

The basal unit is a 3-m thick coral-reef with some coral colonies measuring 1.2 m in length (Fig. 76). A silty matrix is observable between the coral colonies. Clastic sediment input must have been quite high as the corals show a branching – rather than a plate-like growth pattern. The latter would be more common for *Porites sp.* as observed here. Irregular branching trace fossils, possibly belonging to the ichnogenous *Ophiomorpha nodosa* are common within this unit. The trace fossil *Ophiomorpha nodosa* is produced by burrowing endobenthic shrimps lining their cavity walls with fecal balls (Fig. 77).

The basal unit is capped by a massive and grain-supported conglomerate of 1 to 5 m thickness and a maximum grain size in the range of 1 m. This unit is overlain by a 3 m thick silt layer with individual coral reefs containing abundant *Thalassinoides* trace fossils. The uppermost part of the visible profile is a grain-supported conglomerate again of several meters thickness. A thin layer of well-rounded pebbles with abundant

Fig. 77: *Ophiomorpha nodosa* trace fossils in sandy marl deposits at the base of the coral-reef bearing sediments south of Tiwi.

shells makes up the surface of the terrace. Most of it, however, has been mined and is no longer available.

The alternating marine (coral reef) and terrestrial (conglomerates) layers indicate the interplay of eustatic and neotectonic processes. The global (eustatic) sea level shows considerable variations throughout the Quaternary. Regressional phases are characterised by a seaward shift of the coastline whereas transgressions result in the landward shift.

EP 20

Qalhat Fault
Topics: Fault systems, earthquakes, Owen Fracture Zone
Location: UTM 40 Q 749760 2492344 / N 22°31'09 E 59°25'41"

Rating: ☺

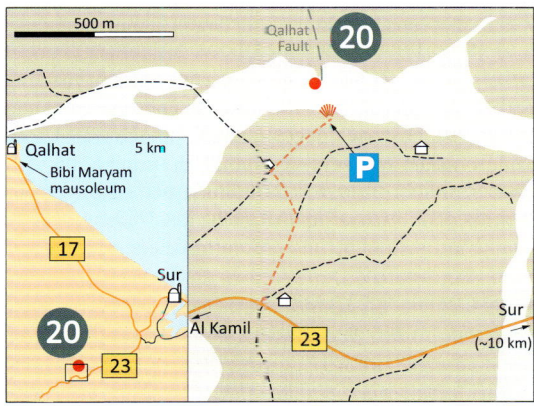

Fig. 78: Location map.

Approach: Follow the highway (#17) from Quriyat to Sur. In Sur, follow road #23 towards the south-west to Al Kamil. At UTM 40 Q 749568 2491484 / N 22°30'41" E 59°25'34" take the unpaved road towards the north and turn left after about 350 m. Follow the road for about 300 m to a radio mast. The outcrop is across the wadi, so after rain it may not be possible to reach the outcrop.

Most of the structural features that are present in the Sur area are linked to the movement of two major fault systems that controlled the deposition of the surrounding rocks and the uplift of the area, at least since the Late Cretaceous. These two fault systems are the Qalhat Fault and the Ja'laan Fault. The Qalhat Fault continues for at least 75 km in roughly a north-south direction, from the southern side of Jebel Qahwan to the sea, close to the location of the Liquefied Natural Gas Company on the coast of Qalhat. They acted initially as normal faults that moved significantly during the Cretaceous Period forming a wide graben in which thick Cretaceous and Paleogene sediments were deposited. This graben represents one of the few areas that received thick clastic deposits in the region during the Cretaceous and Paleogene periods (Al Harthy 2012). The extensional phase of deformation of the Qalhat and Ja'laan faults persisted from Late Cretaceous (Qahlah formation) to the Eocene (Jafnayn formation). Today, the sediments of this basin crop out on the Selma Plateau, which rises up to 2,000 m and represents an uplifted platform of Upper Cretaceous and Cenozoic rocks, bounded by the Ja'alan Fault to the southwest and the Qalhat Fault to the east (Fournier et al. 2006).

A reverse movement of the fault systems has occurred since the Miocene times, about 20 million years ago as evidenced by the syn-sedimentary angular unconformities in the lower Miocene Sur formation in the vicinity of the fault (Wyns et al. 1992). This movement is most likely related to the movement of the Owen Fracture Zone, which represents a transform fault that separates India from Arabia along the Indian Ocean. According to Rodriquez et al. (2011), several episodes of destabilisation occurred during the uplift of the Southern Owen Ridge which began 20 million years ago. The uplift is consistent with early Miocene compressive structures observed inland in the vicinity of the Qalhat Fault (Fig. 79). Historical records show that the well-known trade community that flourished in Qalhat was severely damaged during a major earthquake in the 15th century (Musson 2009). Very few buildings, one of which is the mausoleum of Bibi Maryam in the old town of Qalhat (Fig. 78; UTM 40 Q 743782 2511899 /

Fig. 79: The Qalhat Fault (090/75, normal faulting) between Eocene and Miocene strata.

N 22°41'48" E 59°22'23") survived intact. The earthquake is likely to be related to the movement of the Qalhat Fault because the city was situated almost on top of the fault plane.

EP 21

Sur, mangrove forest, overview point
Note: the road layout will change in the near future as highway construction is in progress at the time of writing.
Topic: Coastal lagoon with a mangrove forest
Location: UTM 40 Q 757779 2496002 / N 22°33'04" E 59°30'24"

Rating: ☺

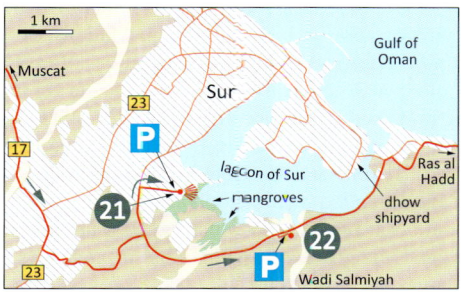

Fig. 80: Location map.

Approach: Follow the highway (road #17) from Muscat through Quriyat to Sur. Proceed toward Sur. After crossing a wadi on the outskirts of the town, you reach the first large roundabout (at UTM 40 Q 755214 2495561 / N 22°32'51" E 59°28'53") with a Shell petrol station on your right. Take the first exit, following the main road towards the south. 700 m after the roundabout turn left and follow the winding road for 2.5 km. Turn left and then after 900 m, take a right turn into a housing complex. After about 700 m you will reach the top of a hill with a nice view of the lagoon. To continue to EP 22, you leave the housing complex and can either choose the direct way back following the road south of the lagoon or continue on the road at the northern rim of the lagoon (see location map, Fig. 80). The road in the north takes you along the lagoon providing scenic views over the lagoon and the city (Fig. 81). You have the option for several stops here to explore the lagoon with its mangroves.

The city of Sur stretches along the coast and covers the area between Wadi Raisah and an irregular shaped lagoon of about 3.5 km by 2.5 km. The old city center is located on

Fig. 81: Overview of the lagoon of Sur with its mangrove forest.

a peninsula stretching into the lagoon. Sheltered from the open sea yet still influenced by tides, mangroves flourish here. They cluster in the area of two wadi-outlets which terminate into the south-western part of the lagoon. Mangroves, here of the species *Avicennia marina* (family Avicenniaceae, Black Mangroves; Pickering & Patzelt 2008), serve an important ecological function as they provide shelter for water birds in general but also for migrating birds and small forms of marine life (Pickering & Patzelt 2008). The scrub-like trees with gnarled trunks reach up to eight meters in height. Mangroves are adapted to the intertidal zone and are therefore able to tolerate a wide range of salinities. Obvious are the so-called pneumatophores, which stick vertically out of the ground. As the typical roots are covered by waterlogged soil, these aerial roots allow the plants to absorb oxygen. Mangroves are vulnerable in Oman as their distribution is limited to only small areas along the coast.

Several lagoons, especially along the rocky coast east of Muscat were destroyed for construction projects. On the other hand, a successful reforestation project has been undertaken by the Ministry of Environment and Climate Affairs and Sultan Qaboos University in the lagoon of Ras al Hadd.

EP 22

Coastal notch close to Sur
Topics: Bio-erosion, uplift
Location: UTM 40 Q 759922 2495750/ N 22°32'55" E 59°31'38"

Rating: ☺☺

For location map see EP 21, Fig. 80.

Approach: Follow the ring road around the city of Sur south of the lagoon to Wadi Salmiyah. Turn right 400 m before the wadi into a narrow road parallel to the ring road which leads directly into the outcrop visible from the street.

Fig. 82: Coastal notch at the mouth of Wad Salmiyah into the lagoon of Sur.

The outcrop is located along the front of a mountain made up of Miocene conglomerates and dolomitised limestone of the Sur formation, overlain by Pleistocene delta deposits (gravel). A coastal notch is obvious within the limestone formation (Fig. 82). This notch is located at 3.7 m above mean high tide level and represents a good sea level index point. The notch formed due to bio-erosion by marine organisms. Shallow holes left behind by *Lithophaga* are obvious below the notch. Other sessile organisms such as oysters can still be observed in situ in sheltered parts of the notch. Radio-carbon dating of these shells produced unreliable and inconclusive dates, probably due to re-crystallisation, as well as exceeding the dating limit. A palaeo-beach can be observed around 100 m to the west of the notch. The sandy beach deposit is also observable as erosional remnants within the notch. Here, samples were collected for optically stimulated luminescence (OSL) dating. The results indicate Marine Isotopic Stage 5 (Eemian interglacial) as the time of beach accumulation (Hoffmann et al. 2013b, Mauz et al. 2015). The date implies that this coastal area is stable in contrast to other crustal blocks which are in uplift as indicated by raised coastal terraces as seen along the coastal stretch from Quriyat to Qualhat (compare EP 14).

EP 23

Carbonatite on the road from Sur to Ras al Hadd
Topics: carbonatite
Location: UTM 40 Q 777402 2490183 / N 22°29'44 E" 59°41'46"

Rating: ☺

Approach: Follow the road from Sur to Ras al Hadd for 21 km. The outcrop is near the road on the eastern slope of the wadi.

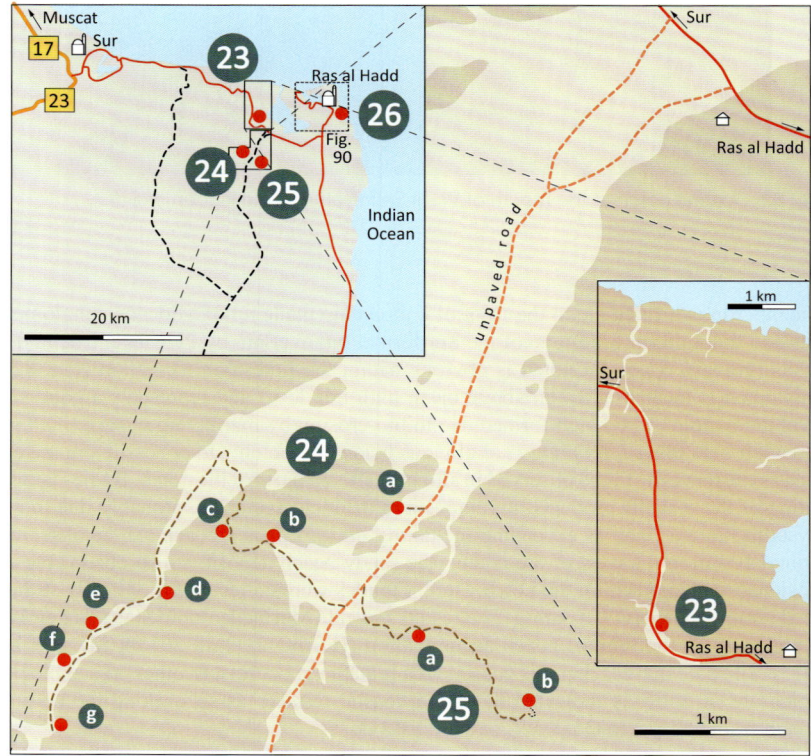

Fig. 83: Location map.

Field sites

Fig. 84: View from the road towards the small hill composed of carbonatite.

Fig. 85: Vein of amethyst crystals exposed on the southern flank of the hill.

This stop is an isolated occurrence of carbonatite, located within the Batain nappe (cf. EP 24). The carbonatite is clearly visible from the road as the dark rocks stand out in the landscape (Fig. 84).

The hill is approximately 30 m across. An amethyst, violet-coloured variety of quartz, vein is observable (Fig. 85). The reader is referred to the work of Nasir et al. (2011) for more information of the occurrences of carbonatite within the Batain nappe.

EP 24

"Mother of all outcrops"
Topic: Radiolarian cherts of the Batain nappe
Location: UTM 40 Q 777763 2486054 / N 22°27'30" E 59°41'56"

Tab. 1: Locations of outcrops (point 24 on the map, Fig. 83):

a	40 Q 777763 2486054	N 22°27'30" E 59°41'56"
b	40 Q 776885 2485822	N 22°27'22" E 59°41'25"
c	40 Q 776674 2485806	N 22°27'22" E 59°41'19"
d	40 Q 776260 2485561	N 22°27'15" E 59°41'03"
e	40 Q 775790 2485265	N 22°27'05" E 59°40'47"
f	40 Q 775625 2485050	N 22°26'58" E 59°40'41"
g	40 Q 775605 2484662	N 22°26'45" E 59°40'40"

Rating: ☺☺☺

For Location map see EP 23, Fig. 83.

Approach: Follow the road from Sur to Ras al Hadd for 24 km. There is a sign to the village at the lagoon called Kwar al Jaramah to the left and 150 m further there is a turnoff onto a dirt road on the right (UTM 40 Q 779367 2489205 / N 22°29'11" E 59°42'55"). Turn right and follow the dirt road for 3.5 km until you reach the first hills on the right hand side. The outcrop 24a is on the right, 100 m from the main track. To reach the other outcrops (b to g) follow a narrow dirt road into a small wadi.

The outcrop described here is one example of the radiolarian cherts that are part of the Late Jurassic to Cretaceous Wahrah formation (location 24a, Fig. 83). There are numerous other outcrops to be found over the whole Batain plain (locations 24b through g, see table and map, Fig. 83). The rocks belong to the Batain nappes, which are allochthonous units. These are made up of sediments and magmatic rocks of Permian to Upper Maastrichtian age that formed in a marine basin of the Neotethys Ocean. During the separation of Gondwana starting in the late Carboniferous to Early Permian, the Batain

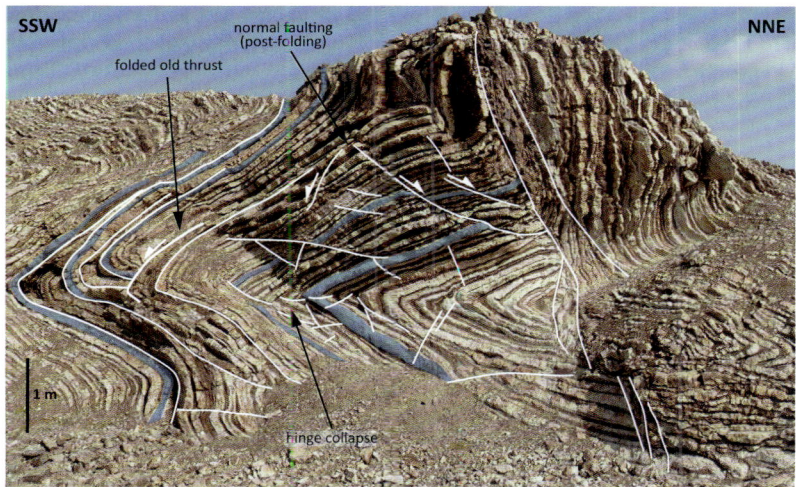

Fig. 86: Radiolarites of the Wahrah formation. Faults and interpretation after Schmid (pers. comm.) and own observations. During a students' excursion in 2010 this outcrop was jokingly named the "mother of all outcrops", a name which has stuck as it has been used in some publications and internet descriptions (e.g., geocaching).

basin opened as a rift basin along the northeastern Oman margin that became oceanic at the Jurassic-Cretaceous boundary (Immenhauser et al. 1998, 2000).

The outcrop (Fig. 86) is 6–7 m high and 25 m in length and consists of brick red and white ribbon-bedded radiolarian cherts (pelagic, deep water sediments), and porcellanite. Porcellanite has a blocky fracture and is an impure chert also containing clay and carbonates. Radiolarian shales and claystones of various colours can also be observed. The individual beds are 5–10 cm thick (Fig. 87). Secondary reducing along the bedding gives the rocks a white appearance from the outside. The rocks are dated by radiolarians as Late Jurassic to Cretaceous in age (Kimmeridgian/Tithonian to Berriasian/Valanginian and from Valanginian to Coniacian).

Radiolarians are holo-planktonic protozoa which are made up of amorphous silica or opal and range in size from hundredths to tenths of millimeters. They accumulated together with clay in the deep sea and diagenetic processes turned the opal into cryptocrystalline quartz. The depth of sedimentation within the ocean basin must have been below the calcite compensation depth as no limestone beds are found within the unit. Therefore, sedimentation must have occurred in water depths below 4,000 to 5,000 m. Due to the high clay content and the absence of turbiditic deposition, it can be deduced that they accumulated at distal positions or parts of the basin that were protected by submarine ridges. The rock sequence exposed shows a well bedded intercalation of red

Fig. 87: Detail of the folded and faulted radiolarites of the Wahrah formation.

and white beds. This intercalation may be explained by changes in sediment influx and diagenetic effects as the silica migrates into the opal-rich chert beds.

Deformation features of the sedimentary layers make this outcrop spectacular. The most characteristic structural features are impressive folds and faults. The folds are mostly reclined chevron folds (Figs. 86, 87). They are tight with moderately dipping axial planes. Hinge collapse can frequently be observed within the outcrop. While some late normal faults cut across the folds and clearly post-date folding, some of the faults represent minor thrusts that formed before and during folding. In general, the Wahrah formation shows large-scale folds whose fold axes trend E-W.

At the time of the Cretaceous to Tertiary transition, west-northwest directed shortening led to the closure of the Batain basin and detachment of the Batain basin from the continental slope, resulting in the formation of the Batain nappes (Peters et al. 2001, Schreurs & Immenhauser 1999). These nappes were thrust upon the eastern Oman margin and were heavily deformed during their obduction. It has to be noted, that the obduction of the Batain nappes occurred approximately 15–20 million years after the obduction of the Samail Ophiolite. During the Tertiary, tectonic processes following the collision of the Arabian with the Eurasian plate such as the opening of the Red Sea and the Gulf of Aden, affected the obducted nappes by further shortening.

EP 25

Manganese Pit
Note: The last 1.6 km to this outcrop is the old access road to the former mining site and abandoned nowadays. The track may be difficult to navigate as several gullies cross it.
Topic: Manganese ore, radiolarian chert
Location: UTM 40 Q 778589 2484806 / N 22°26'49" E 59°42'24"

Rating: ☺☺

Location map: see EP 23, Fig. 83.

Approach: Follow the road from Sur to Ras al Hadd for 24 km. There is a sign to the village at the lagoon called Kwar al Jaramah to the left and 150 m further there is a turnoff onto a dirt road on the right (UTM 40 Q 779367 2489205 / N 22°29'11" E 59°42'55"). Turn right and follow the dirt road for 4.2 km until you reach the first hills. You will pass EP 24 after 3.5 km on this route. Leave the main dirt road to the left at (UTM 40 Q 777504 2485565 / N 22°27'14" E 59°41'47"). Follow this small track for another 1.6 km until you reach the outcrop.

The outcrop (coordinates are of EP 25b in Fig. 83) is part of the Late Jurassic to Cretaceous Wahrah formation as described in more detail in EP 24 (see also Kickmaier &

Fig. 88: Abandonded mine pit of manganese ore in the Wahrah formation.

Fig. 89: Manganese enriched layers in radiolarian cherts of the Wahrah formation.

Peters 1990). The site is an abandoned manganese mine. Open pit mining was carried out here for a couple of months only. The mining activities resulted in a trench of about 200 m long, 10 m deep and 15–20 m wide (Fig. 88). Several tailings are evident in the locality. Furthermore, slabs of ore can be found on the former ore collection sites in the vicinity (UTM 40 Q 778286 2484783 / N 22°26'48" E 59°42'13"). The mine is located in a NW-SE trending stratiform manganese horizon. Black cherts enriched in manganese have individual thickness of up to 50 cm and stand out as massive beds in the mining pit but also at several locations in the surroundings of the pit (e.g., location 25a in Fig. 83). The host rock is radiolarian chert as described in EP 24 (Fig. 89). Several occurrences of these metalliferous minerals can be found in the area. Most of these have recently been exploited and the mining sites are abandoned. The age of the manganese enrichments is Berriasian to Early Valanginian based on analyses of the radiolarian assemblage.

EP 26

Ras al Hadd
Topic: Early Bronze Age settlement, extreme-wave events
Location: UTM 40 Q 788690 2490588 / N 22°29'50" E 59°48'21"
Rating: ☺☺

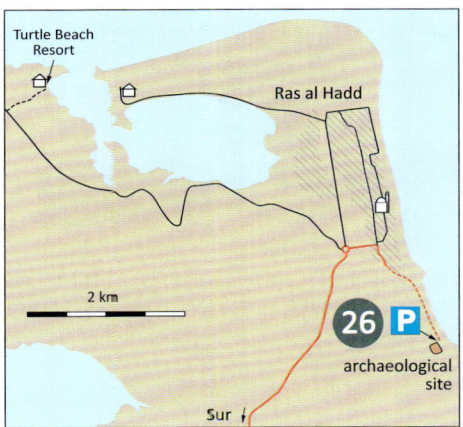

Fig. 90: Location map

Approach: Take the road east towards the coast at the roundabout when entering Ras al Hadd. Turn right at the end and follow the dirt track through a large gate. The gate marks the border of the turtle sanctuary and is closed at night. Follow any of the dirt tracks by the shore parallel to the archaeological site HD-6 at the southern end of the beach.

The coastal area of Ras al Hadd is very rich in archaeological remains dating from the 3rd millennium BCE and later (Cleuziou & Tosi 2007c, Hannss 1998). The archaeological site HD-6 was discovered in 1986 and has been subject to archaeological investigations from 1996 to the present (Azzara 2009, Cartwright & Glover 2002, Hilbert & Azzara 2011). The site covers a circular area of approximately 10 000 m² and is elevated by a couple of decimeters above the surrounding landscape. The mount appears to have been densely settled and three main periods of occupation have been identified.

The main phase is represented by sand and mud brick structures, and marks the very beginning of the Early Bronze Age. The absolute age for this period is given as 3100–2700 BCE. A total number of fourteen buildings surrounded by a stone wall were dug out (Fig. 91). The walls of the adobe structures were made up of standardised mud

Fig. 91: View into the excavation of HD-6 (February 2013). The foundations of the houses made up of clay bricks as well as the stone wall in the back are visible. For scale, the girl's height is 0.85 m.

bricks of 55 × 32 × 8 cm which were joined by 1–2 cm layer of mortar. Smoothed clay was used for the floors, and square tiles of 35 × 35 cm have also been taken note of. The houses have 3 to 6 rooms where the individual rooms are rectangular and rather small, only a few square meters each. Archaeological evidence suggest a comparatively large population of around 150–200 inhabitants occupying the site over several centuries, however probably only seasonally during the winter months. The function of the rooms is not unambiguously clear. They were probably used as storage facilities or for the manufacture of domestic supplies and ornamental goods. Fireplaces are common inside and outside the buildings.

The economy was dominated by fishing as evidenced by numerous fishing related artefacts such as ropes, nets, net sinkers and copper fishing hooks. The fishing gear was sophisticated so that even large prey such as dolphins was caught. Cold hammering of metal objects appears to have been a common activity. There is lack of evidence, e.g., slag deposits, that copper was smelted on site; it therefore had to have been imported from elsewhere. The copper was used for pins, knives, chisels, and many other small tools. There are plenty of remains indicating the manufacturing of beads.

Sedimentological evidence exists that the site was inundated by an extreme-wave event around 2500 ± 200 BCE (Hoffmann et al. 2015). The causative process is either storm or tsunami. The interpretation as tsunami inundation is favored as no overwash deposits were observed in the last 2 cyclone events, in 2007 and 2010 respectively.

EP 27

Salt lake near Al Ruwais
Topic: Sabkha, salt evaporation
Location: UTM 40 Q 781084 2454673 / N 22°10'28" E 59°43'32"

Rating: ☺

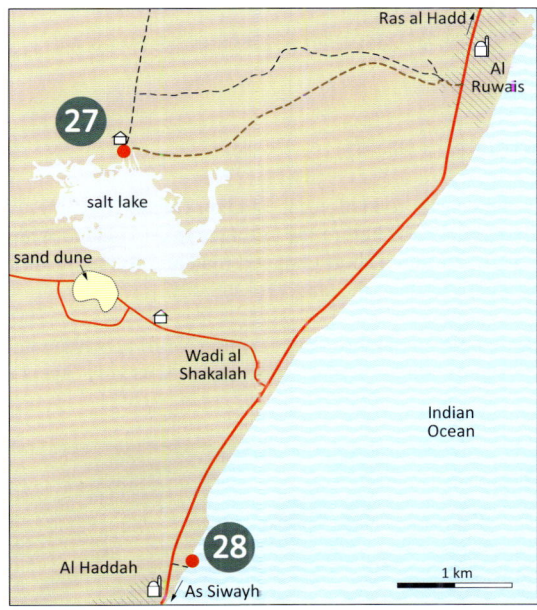

Fig. 92: Location map.

Approach: Leave the coastal road along the Indian Ocean coastline from Ras al Hadd towards the south in Al Ruwais. Turn right onto an unpaved road and follow this road along the Wadi al Shakalah for about 4 km to the west.

The wadis that drain Jebel Khamis with its basement outcrops 30 km west of the coast (compare EP 31) often have no open connection to the sea, as the gradient is very low and a range of low hills (Batain nappes or Tertiary rocks) forms barriers. Sabkha deposits accumulate in topographic depressions. Here clay and salt are deposited.

The stop is on the northern side of one of the larger sabkhas with a size of roughly 2 × 1 km. The size of the sabkha, or coastal saline depression, depends on the amount of rainfall, which in general is only around 50 mm per year. Rainfall events are sporadic and episodic but can be very intense, leading to an infill into the lake. Small scale salt-mining activities can be observed around the shores of the lake.

EP 28

Al Haddah
Topic: Oyster shells at a coastal cliff
Location: UTM 40 Q 781495 2450172 / N 22°08'02" E 59°43'44"

Rating: ☺

For location map see EP 27, Fig. 92.

Approach: Leave the coastal road along the Indian Ocean coastline from Ras al Hadd towards the south about 400 m before the entrance to Al Haddah which is also named Bandah Jadidah. Turn left into a short unpaved road and follow it for about 100 m to a small shelter. A footpath leads to the small coastal cliff.

The outcrop is located directly at the coast, forming a cliff approximately 5 m high. The sequence is composed of conglomerates, sandy as well as silty layers of inferred Qua-

Fig. 93: Large locular carbonate shells at the coastal cliff near Al Haddah.

Fig. 94: *Crassostrea gryphoides*. The upper picture shows the outer appearance of the shell. The lower picture shows a section through the shell illustrating the internal structure.

ternary age. These deposits were laid down in a near shore coastal environment. They most probably represent beach deposits that formed during the last interglacial period (Eemian Interglacial, Marine Isotope Stage 5e). The Quaternary rocks discordantly overlie a series of vertically inclined layers of shelf sediments with alternating reddish radiolarian mudstones and grey pelagic limestone which belong to the Triassic Sal Formation of the Batain basin (Peters et al. 2001).

Carbonate shells of up to 30 cm in length are abundant in the upper layers of the Quaternary sequence (Figs. 93 and 94). The round, elongated tube-like and locular shells are remnants of oysters (*Crassostrea gryphoides*). They are preserved with one valve only, indicating that they accumulated along the palaeo-beach and were not preserved in-situ.

EP 29

As Siwayh shell midden
Note: The middens are protected, so refrain from taking any shells along as souvenirs.
Topic: Prehistoric shell midden
Location: UTM 40 Q 777651 2445323 / N 22°05'26" E 59°41'27"

Rating: ☺

Approach: Follow the coastal road along the Indian Ocean coastline from Ras al Hadd towards the south for around 54 km until you enter As Siwayh. Tall piles of broken shells can be found directly beside the road.

Fig. 95: Location map.

The coastline is dotted with these shell middens. The more than 50 sites that are known between Ras al Jinz in the north and Asylah in the South, provide evidence for prehistoric subsistence economies (Biagi 1994a, Berger et al. 2013). The middens appear as circular structures which may be up to several meters high and tens of meters in diameter. The material found in the shell middens comprises shells of various edible species including oysters and other bivalves as well as gastropods. Especially the gastropod shells (e.g. *Terebralia palustris)* are fragmented, indicating the mechanical opening of the shell to access the soft part of the animal for consumption. Besides the dominating shells, remains of bones, fishes and crabs are common. Fishing equipment such as net sinkers, mother-of-pearl shellhooks and line-weights are documented as are many types of stone tools (Berger et. al. 2013). There is no evidence for agriculture, and herding activities were only marginal.

The material in the middens was accumulated by Neolithic fisher communities that lived in the coastal environment from the 6th to the 4th millennium BCE. The material indicates that lagoon and mangrove environments were exploited. This in turn allows the conclusion to be drawn that the climate must have been wetter than today. The shell middens are protected archaeological sites. Hence disturbances, collection of material, etc. should be avoided.

EP 30

Beach 5 km south of As Siwayh
Topic: Kimberlite, carbonatite, radiolarite
Location: UTM 40 Q 775728 2438440 / N 22°01'44" E 59°40'15"

Rating: ☺☺

For location map see EP 29, Fig. 95.

Approach: From Sur take road #23 to Al Kamil and turn left on road #35 to Jalan Bani Buhassan. Drive through Jalan Bani Buhassan and follow road #35 for about 35 km. Turn left into the road to Asylah and there turn left again to the north to As Siwayh (see location map at EP 29, Fig. 95). After 10 km (at UTM 40 Q 779367 2489205 / N 22°02'43" E 59°39'54") turn right onto a short paved road leading to the coast and follow an unpaved sandy road behind a narrow ridge for about 2 km to the south. The outcrop is at the beach.

The outcrops are situated on the beach about 5 km south of the village of As Siwayh. There are only a few blocks of kimberlite, several meters in size (Fig. 96) at the northern part of the beach. The rock is mainly composed of a dark groundmass with xenolithic components of carbonatite, chromite, gabbro and peridotitic material. The components are up to 30 cm in diameter. At some places the rocks contain round lapilli and

Fig. 96: Kimberlite blocks at the beach 5 km south of As Siwayh.

Fig. 97: Concentric lapilli and ultramafic components of the kimberlite at the beach south of As Siwayh.

cm-sized phenocrysts and xenoliths. The lapilli are about 2 cm in size with concentric layers of mineral growth around a core particle (Fig. 97).

Kimberlite is quite a rare volcanic rock that formed in vertical pipes and igneous dikes and sills. It often contains rock material derived from depths greater than any other igneous rock type. It is named after the town of Kimberley in South Africa, which is famous for prolific diamond mining in the past. Kimberlite ascends from the mantle very quickly, typically within a few hours. Frequently found components are serpentinised olivine, phlogopite, carbonatic minerals as well as garnet, enstatite and chrome minerals (Vinx 2005).

In this particular outcrop, two rock varieties can be discerned: a heterolithic breccia and a lapilli tuff. On a microscopic scale the matrix of the heterolithic breccia also contains components of phlogopite peridotite, chromite grains, carbonatite and trachytic basalt (Peters et al. 2001, see also Nasir et al. 2011, for detailed information on the petrogenesis). However, diamonds were not found at this location.

The concentric lapilli that dominate in parts of the kimberlite develop during the ascent of the magma. The higher the magma gets, the more dissolved gas passes into the gas phase and expands due to pressure reduction. At a certain point turbulent currents occur in the gassy magma stream. Under these conditions lapilli can form around core particles. The material that may become a core can be early crystallised minerals or rock fragments from the wall of the pipe.

The kimberlite pipe of As Siwayh has a diameter of 200 to 300 m (Nasir et al. 2008). It intruded into radiolarian cherts of the Wahrah formation (Batain nappes) and was dated using several radiometric methods. Ar/Ar dating of the phlogopites proved an age of 150 Ma (Late Jurassic; Peters et al. 2001). U-Pb zircon model age obtained by Nasir et al. (2008) revealed an age of 137 ± 1 Ma (Early Cretaceous).

At the southern end of the beach, the radiolarian cherts are exposed with a nice fold structure. The radiolarian cherts are overlain by conglomerates which contain a block of carbonatite of unknown origin which is several meters thick.

EP 31

Basement rocks in Jalan Bani Buhassan
Topic: Precambrian plutonites, granite, pegmatitic dikes
Location: UTM 40 Q 744125 2449325 / N 22°07'54" E 59°22'00"

Rating: ☺

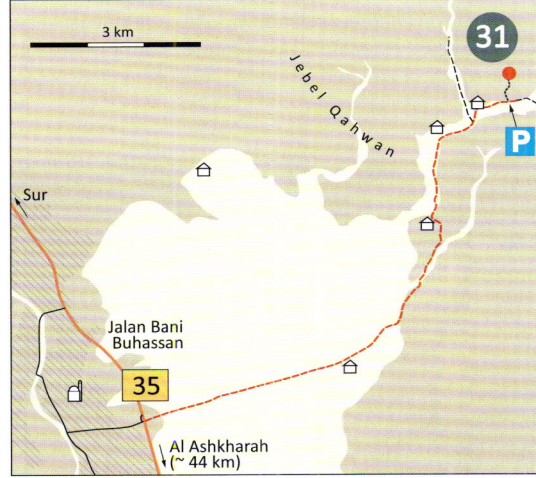

Fig. 98: Location map.

Approach: Drive from Muscat on the road to Ibra, then follow the road from Ibra to Sur. An alternative would be to drive from Muscat to Sur via the dual carriageway coastal road, then follow the road from Sur to Ibra. Coming from Muscat, take a right turn in Al Kamil (UTM 40 Q 727161 2459293 / N22°13'27" E 59°12'13"). This would be a left turn if you were driving from Sur to Ibra. Continue driving to Jalan Bani Buhassan (road #35). Turn left to Jebel Qahwan at UTM 40 Q 738141 2442566 / N 22°04'18" E 59°18'28" and drive straight to the end of the blacktop road; then follow the graded road. At this road, you are surrounded by crystalline, mainly igneous rocks that represent the basement of the Arabian Plate. Leave the road at UTM 40 Q 743511 2448350 / N 22°07'23" E 59°21'38" about 400 m before entering a little village and follow the wadi towards the north for about 1.6 km. The outcrop is on the left hand side at the slope of the wadi. The point itself marks the top of the hill.

Jebel Qahwan represents the extreme eastern end of the Al Hajar Mountains of northern Oman, which extend from the Governorate of Musandam to Ras al Hadd for about 700 km. This mountain is also part of the uplifted eastern margin of Oman. It is composed of gneisses, amphibolites and granites. These crystalline rocks represent outcrops of the basement of the Arabian Plate. They formed more than 800 million years ago when a series of terranes or micro-continents accreted at the northeastern part of Africa during the Proterozoic. The age of the rocks is determined by radiometric age dating. Similar outcrops can also be visited along the central and southern parts of the eastern coast of Oman, in the wilayats (districts) of Mahout and Mirbat, respectively.

The crystalline light-coloured basement rocks are often cut by younger dark basaltic dikes. Across Jebel Qahwan, these dikes are oriented in an east-west direction, indicating a north-south directed extensional event during the Proterozoic. The whole sequence of basement rocks is covered by Late Cretaceous and Paleogene sediments.

Fig. 99: Pegmatitic dike in plutonic rocks of the Ja'alan batholith.

The allochthonous units of the Hawasina deep-oceanic sediments and the Samail Ophiolite thrust sheets are missing here. The uplift and inversion of Jebel Qahwan most likely occurred after the Middle Eocene during a compressional event resulting in NS-trending anticlines (Filbrandt et al. 1990). The Jebel Qahwan region is important because it exhibits structures that flank the oil-bearing sedimentary rocks to the south.

The outcrop exposes granites of the Ja'alan batholith (Roger et al. 1991) cut by pegmatitic dikes with large feldspar crystals (Fig. 99).

EP 32

Wahiba Sand
Topics: Desert formation, aeolianites, Quaternary climate change
Note: The cemented dune deposits of the Wahiba Sands are described as the most extensive continuous deposit of aeolianites in the world
Location: UTM 40 Q 675750 2484845 / N 22°27'39" E 58°42'29"

Rating: ☺☺

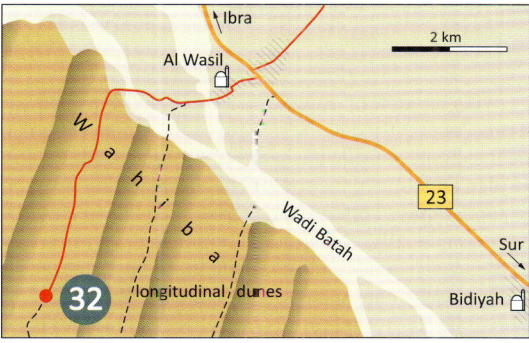

Fig. 100: Location map.

Approach: Leave the main road #23 in Al Wasil towards the west and follow the paved road into the longitudinal dunes of the Wahiba Sands. After crossing two broad wadis, you will reach the red-coloured dunes in about 3.5 km. The paved road continues for more than 4 km into the dunes.

The Wahiba Sand Sea is bordered by Wadi Batah to the north and to the east where fluvial activity truncates the aeolian deposits. The western limit is defined by Wadi Andam, the Indian Ocean borders to the south. The adjacent shelf is only 80 km wide.

Fig. 101: The Wahiba Sand with small recently formed barchan dunes on top of the old longitudinal dunes. The location is near Al Wasil.

The sand desert covers an area of approximately 16,000 km^2 and extends for 100 to 200 km from the present coastline to the margin of the Oman Mountains. The sand sea fills an area of local tectonic subsidence between the Oman Mountains in the north and the Huqf-Jebel Ja'alan anticline in the south. The most striking geomorphologic feature is the mega-dune topography. This is characteristic for the so called High Sands. The Low Sands are stratigraphically younger and dominated by linear and barchan dunes (see EP 41). High transverse dunes and aeolianite make up the coastal area, whereas the Al Jabin Plateau is entirely composed of cemented dune deposits.

The mega-dunes of the High Sands are asymmetric linear ridges up to 70 m high (Fig. 101). Their orientation is approximately south to north (NE/SW and NNE/SSW) and the dunes have a crestal spacing of 1 to 2 km. The extension of the individual linear dunes exceeds 100 km in cases. The thickness of aeolian sediments is up to 150 m. The cemented dune deposits are described as the most extensive continuous deposit of aeolianites in the world (see also EP 43). Limnic deposits can be found in the swales of the dunes, indicating small seasonal palaeo-lakes (Preusser et al. 2002, 2005, Radies et al. 2005).

Composition of the sands

The composition of the aeolian sediments is dominated by a high amount of allo-chemical fragments. This holds especially true for the aeolianites, where up to 90 % of the sand-sized particles are made up of shell fragments, foraminifera and peloids (Gardner 1988). These sandstones are weakly cemented, usually by low-Mg calcite and locally by halite. The large linear dunes have a higher concentration of terrigenous material such as silicate and rock fragments. Opaque minerals derived from the ophiolite are uncommon; an exception is the northeastern corner of the sand sea. Reworking of the older aeolianites is reported. The general appearance of the sand is red. This colour is partly of primary origin, where the sand particles are terrigenous red cherts and partly of secondary origin where a red iron oxide coating is observed on the grain surfaces. Additionally, finer ferruginous sediments (silt and clay) are observed.

Quaternary climate changes

The Wahiba Sand Sea is climatically influenced by the continental interior of Arabia as well as the Indian Ocean and situated at the northern limit of the area currently affected by the Indian Ocean Monsoon (Glennie & Singhvi 2002). The modern climate of the region is hyper-arid with precipitation of less than 100 mm per year. Occasional rainfall is associated with tropical cyclones during the winter months. Substantial climate variability is recorded throughout the Quaternary. There is a correlation of high latitude glaciations, low global sea level and decreased monsoon activity. Aeolian processes were most effective and pronounced during periods of high latitude glaciations. The dune sediments were deflated from the shelf and wadi beds. These arid phases were characterised by less vegetation cover and hence increased sediment availability. Higher rainfall and fluvial activity are recorded for the interglacial periods. The general transport direction of the Aeolian sediments is from south to north. The recent topography is a Pleistocene relict.

Dating of the aeolian deposits is done by utilising optically stimulated luminescence (OSL) methods (Preusser et al. 2002, 2005, Radies et al. 2005). The oldest dates obtained on the aeolianites indicate an initial dune formation around 140,000 to 160,000 BP. Luminescence dating further proved that the large linear dunes were established during 120,000 to 100,000 BP. Dune activity in the High Sands is also recorded in the periods 71,000–57,000 BP, during the last glacial maximum (24,000–11,500 BP) and to a lesser extent during the latest Holocene. A lack of synchronicity in aeolian activity is observed in the southern and northern part of the Wahiba Sands.

Seasonal and even permanent lakes existed in the early Holocene (ca. 9,300–5,500 BP). This Early Holocene wet period is also recorded in other palaeo-climate records, such as speleothems throughout southern Arabia.

EP 33

5 o'clock Moho
Note: It is worth aiming to be at this site in the late afternoon to get the best light and clearest view of the line that divides the lithospheric mantle and the oceanic crust of the Samail Ophiolite.
Topics: Mohorovičić discontinuity viewpoint
Location: UTM 40 Q 629425 2536749/ N 22°56'02" E 58°15'44"

Rating: ☺☺

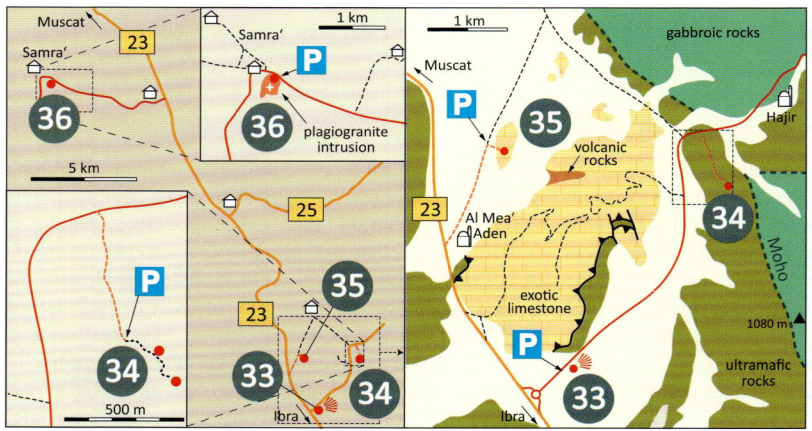

Fig. 102: Location map.

Approach: Follow the main road #23 from Ibra to Muscat. About 2.2 km after the junction with road #27 (not shown on location map, Fig. 102) turn right (at UTM 40 Q 628951 2536306 / N 22°55'47" E 58°15'27") into a roundabout and follow the paved road for a few hundred meters. Stop at the shoulder. When driving from Muscat to Ibra, there is no exit. In this case you have to do a U-turn to face the opposite direction at the nearest opportunity.

At this stop you have to look towards the east to a mountain which rises about 400 m above the plain (Fig. 103). You will have a fascinating distal view of the clear boundary line between the lithospheric mantle and the oceanic crust of the Samail Ophiolite. The best time to look at the outcrop is at 5 o'clock in the afternoon just before sunset. Hence the term "5 o'clock Moho" was introduced by Adolphe Nicolas and co-workers.

Fig. 103: View of the boundary between lithospheric mantle and oceanic crust, the Moho, from the distance at viewpoint EP 33 (see Fig. 102).

The lithospheric mantle is defined by ultramafic rocks, mainly composed of harzburgite and dunite at this location. The upper part of the mountain consists of gabbroic rocks which represent the lowest part of the oceanic crust (Fig. 103).

EP 34

Chromite pit
Topics: Podiform chromite, ultramafic rocks
Location: UTM 40 Q 631403 2539042 / N 22°57'15" E 58°16'54"

Rating: ☺

Location map see EP 33, Fig. 102.

Approach: Follow the main road #23 from Ibra to Muscat. About 2.2 km after the junction with road 27 (not shown on location map, Fig. 102) turn right (at UTM 40 Q 628951 2536306 / N 22°55'47" E 58°15'27") into a roundabout and follow the paved road towards the north-east. When driving from Muscat to Ibra, there is no exit. In this case you have to do a U-turn to face the opposite direction at the nearest opportunity. After about 4.5 km on the paved road, turn right behind a right hand bend onto an unpaved road and follow this road towards the south for 800 m uphill.

Fig. 104: Chromite vein in ultramafic rocks of the Samail Ophiolite.

The outcrop is in the middle of a pit within a series of ultramafic rocks of the Samail Ophiolite. The chromite body is quite inconspicuous as it is similar in colour and appearance to the ultramafic rocks. The ore body is exploited but tailings left by the mining activities and even some in-situ remains of the chromite ore can be found at this location (Fig. 104). The ultramafic rocks are substantially altered into serpentinite. At some places nice serpentine minerals can be observed (Fig. 105).

Accumulations of chromite occur as stratiform layers in cumulates above the harzburgites (Fig. 18), as lenses or pods and veins within dunites in the upper mantle within 1000 m of the Moho where the chromite ore is typically associated with dunitic metasomatism of the harzburgitic host rock (Rollinson 2005). Chromium oxide (Cr_2O_3) content of ores typically exploited ranges from 30 wt. % to 42 wt. % and even up to 54 wt. % (Rajendran et al. 2012).

Mining activities are restricted to open pit mining whereby obvious, superficial outcrops are exploited. The level of automation is low. Operationally a load and carry method is used to transport material. Typically, excavators and bulldozers would work either independently or in conjunction with heavy trucks to transport the material to a concentration point for sorting and screening.

Fig. 105: Serpentine minerals in altered ultramafic rocks of the Samail Ophiolite.

EP 35

Oman exotics
Topics: Reef limestone, rhyolitic volcanic rocks
Location: UTM 40 Q 630365 2539183 / N 22°57'20" E 58°16'17"

Rating: ☺

Location map see EP 33, Fig. 102.

Approach: Follow the main road #23 from Ibra to Muscat. About 4.4 km after the junction with road 27 (not shown on location map) turn right at the little village Al Mea'Aden (at UTM 40 Q 627859 2538252 / N 22°56'50" E 58°14'49") onto a paved road towards the north-east. After about 1.3 km leave the car at the road and walk about 150 m to the clearly visible outcrop.

The outcrop described here is at the base of a light-coloured hill of 200 m high and ca. 1.0 × 3.0 km across. It stands out in a landscape dominated by the brown rocks forming jagged hills of the mantle section of the Samail Ophiolithe. Mining activities give the mountain a ruined appearance. The limestone is mainly used as granulates for the production of terrazzo tiles.

The sedimentary formations exposed here in the geological window of Wadi Musfa are part of thrust sheets of the Hawasina nappes. Structurally they appear as an anticlinal upwarp trending N 150–160° and tectonically imbricated with serpentinised peridotite. The dominating lithology is white reefal limestone associated with red radiolarian chert which rests on a rhyolitic volcanic substrate. The stratigraphy is Late Permian to Early Triassic and rock formation took place in a proximal subtidal environment. The limestone is characterised as a breccia at the base and a volcanosedimentary, chalcedony matrix is described by de Gramont et al. (1986) indicating contemporaneous volcanism. The same authors describe fossil remains of bryozoans, branching corals, sponges, brachiopods, echinoderm fragments, algae and foraminifera. Isolated hills in the landscape are made up of yellow lithoclastic limestone. Some of those are karstified and weathered to peculiar formations. One of those appears like a giant skull (at UTM 40Q 628509 2539425 / N22°57'29" E58°15'12"; Fig. 106). The volcanic basement of the reef is exposed a few hundred meters to the east of the giant skull (see location map). These volcanites may be interpreted to have

Fig. 106: Karstified and weathered limestone of the Late Permian to Early Triassic in form of a giant skull.

formed as a seamount which led to the development of reefal limestones on top and breccias at the sides. The formation represents one of the so-called "Oman exotics" (cf. Searle & Graham 1982).

The Oman exotic blocks were once isolated klippen or atolls (see EP 81 for another example). They were emplaced on the northern part of Oman during the Late Cretaceous obduction process and often found within a complicated assemblage of thrust sheets of deep oceanic sediments, usually thin-bedded shales and cherts, as well as volcanic and mélange units that tectonically underlie the ophiolite thrust sheets. They consist of white limestone that crop out within thrust faults beneath the Samail Ophiolite in the Al Hajar Mountains. They were termed "exotic" because they look out of place in their surrounding of dark-coloured deep-oceanic sediments and Samail Ophiolite rocks. These exotic blocks range in size from a boulder size to more than 1,000 m thick. The Late Permian exotics are of dominantly reef and forereef facies, whereas Triassic exotics are more typically of back-reef and lagoonal facies. They are commonly associated with a substrate of alkaline and transitional tholeiitic basalts and are interpreted as a series of reef-associated carbonate build-ups deposited in part on oceanic islands or seamounts in the Neotethys Ocean (see Fig. 15), close to the site of initial rifting of the Oman continental margin (Searle & Graham 1982). Some of the exotic blocks may have slumped down from large underwater cliffs before their obduction.

EP 36

Plagiogranite
Topics: Plagiogranite intrusion
Note: the highway from Bidbid to Ibra is under construction at the time of writing, changes to the road layout are to be expected.
Location: UTM 40 Q 616005 2552890 / N 23°04'50" E 58°07'57"

Rating: ☺

For location map see EP 33, Fig. 102.

Approach: Leave the main road #23 from Ibra to Muscat 7 km after the junction with road # 25, turning left into a paved road (at UTM 40 Q 621596 2552020 / N 23°04'20" E 58°11'13"). Follow this road for about 6 km. The outcrop is at a light coloured hill on the left hand side of the road.

Plagiogranites usually form small plutonic bodies that developed either at oceanic spreading centers by extreme fractional crystallisation from a mantle melt or by partial melting of hydrated mafic (gabbroic) crust. The formation of SiO_2-rich melts within the oceanic crust is explained by partial melting of gabbroic rocks affected by a fluid phase mainly composed of H_2O from seawater (olivine + clinopyroxene + plagioclase + H_2O → orthopyroxene + partial melt + plagioclase). Based on the geochemistry of oxygen

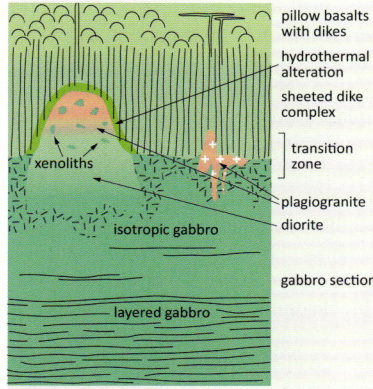

Fig. 107: Schematic cross section of oceanic crust with plagiogranite intrusions in the dike-gabbro transition zone (modified after Grimes et al. 2013).

Fig. 108: Light coloured plagiogranite with xenoliths (dark rock particle in the middle part of the photograph) of mafic crust from the upper part of the Samail Ophiolite sequence.

isotopes (primary magmatic vs. seawater-contaminated $\delta^{18}O$ values) the plagiogranites of Oman are attributed to partial melting (Grimes et al. 2013).

The International Union of Geosciences defines plagiogranites as light-coloured plutonic rocks with mostly plagioclase and quartz and less than 10 % of mafic minerals. The content of K_2O is less than 0.2 %, thus potassic feldspar (orthoclase) does practically not occur. The composition of plagioclase is dominated by albite. Synonyms of plagiogranite are trondhjemite and tonalite which may be subdivided by the albite-anorthite relation of the plagioclase. The Oman plagiogranites are composed of plagioclase and quartz with only minor amounts of ferromagnesian minerals such as amphibole or biotite. Their formation took place in the transition zone from gabbros to sheeted dikes (Fig. 107). The abundant xenoliths of mafic rocks (Fig. 108) indicate intrusion into the already crystallised mafic crust. The age of the plagiogranite is similar to other rocks of the Samail Ophiolite. Warren et al. (2005) determined a U-Pb age from zircons of the plagiogranites with a mean value of 95.3 ± 0.2 Ma.

EP 37

Metamorphic sole near Bidbid
Topic: Metamorphic sole of the Samail Ophiolite
Location: UTM 40 Q 616852 2590519 / N 23°25'13" E 58°08'37"

Rating: ☺☺

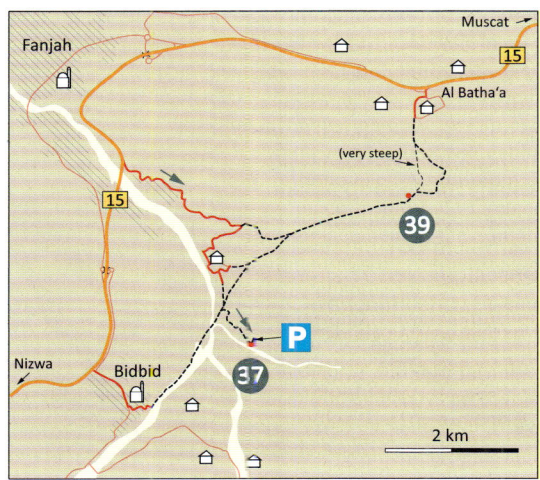

Fig. 109: Location map.

Approach: Leave the main road #15 from Nizwa to Muscat in Fanjah immediately after crossing the bridge over the wadi towards the right. Follow the winding paved road along the intensely red-coloured mountains for about 2.5 km. Follow the road downhill to a little village on the bank of Wadi Fanja. Drive through the village and continue on an unpaved road for about 500 m. You will reach a broad unpaved road which is a service road parallel to the oil pipeline which leads down into the wadi. A dirt track to the outcrop starts on the opposite side of the broad unpaved road. You will reach the outcrop after about 1 km, arriving at a little creek.

If you are coming from the direction of Nizwa it is also possible to leave the main road #15 in Bidbid. Pass through the village of Bidbid and follow the straight unpaved road which is the service road parallel to the oil pipeline towards north-east into the wadi. Turn right onto the dirt track and reach the outcrop after about 750 m (see above).

The metamorphic sole is an approximately 150–200 m thick unit of tectonites at the base of the Samail Ophiolite nappe. It was formed at the early stages of intra-oceanic thrusting and led to high-temperature amphibolite and locally granulite facies recrystallisation of basaltic/gabbroic rocks of the underthrust ophiolitic and sedimentary rocks. The metamorphic sole was formed in the intra-oceanic thrust zone near the mid-oceanic ridge and thermally overprint by the heat from the still hot mantle of the overriding Samail Ophiolite (Tethyan type of ophiolites; see Fig. 17). This so-called "ironing ef-

Fig. 110: Folded amphibolites of the metamorphic sole at the base of the Samail Ophiolite nappe.

fect" (see Nicolas 2016), is only possible if thrusting occurred at very high temperatures near the ridge axis. This interpretation is consistent with U-Pb ages from ophiolitic gabbros, trondhjemites and tonalites that yield only about 300,000 years between the last magmatic events related to ridge activity and early thrusting confined by metamorphic melt products in the metamorphic sole (Warren et al. 2005, Rioux et al. 2013). This model suggests that the metamorphic sole was intra-oceanic and obduction, in the sense of emplacement of oceanic lithosphere on a continental lithosphere, followed, when the overriding Samail Ophiolitic nappe reached the passive margin of Arabia.

At location EP 37 (Fig. 109) a series of meta-sediments and amphibolites represents the metamorphic sole of the Samail Ophiolite. It is squeezed between deformed rocks of the Hawasina nappe with turbiditic deposits in the north-east and peridotites of the Samail Ophiolite nappe in the south-east, on the opposing side of the wadi. Quartzites, muscovite-bearing quartz-schists, mica- and chlorite-schists, and locally crystalline limestones represent the metamorphosed ocean sediments. Dark green striped amphibolites (Fig. 110), with alternating amphibole- and plagioclase-rich layers, are attributed to the metamorphosed oceanic crust overridden by the Samail Ophiolite. Actinolite, chlorite and epidote indicate late retrogression to greenschist metamorphic conditions.

EP 38

Listwanite
Topic: Carbonatised ultramafic rock, listwanite
Location: UTM 40 Q 620847 2584379 / N 23°21'52" E 58°10'56"

Rating: ☺

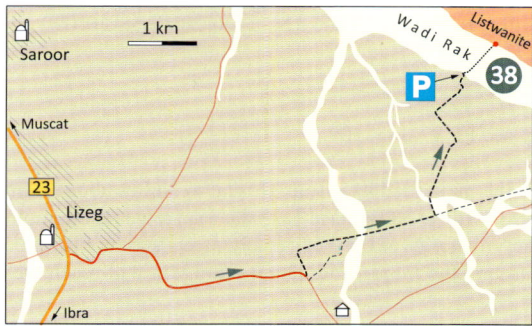

Fig. 111: Location map.

Approach: Leave the main road #23 from Muscat to Ibra in Lizeg about 10 km after the turnoff from road #15. Turn left in Lizeg and follow the paved road towards east

Fig. 112: Panorama view of the outcrop of iron-rich carbonatised ultramafic rocks (listwanite).

traight ahead for 4 km. Then turn left into a dirt road (at UTM 40 Q 620057 2581626 / N 23°20'23" E 58°10'27") and turn right after 400 m, continuing down the road for 2 km. A narrow dirt road towards north leads to the edge of the broad gravel-filled Wadi Rak. Leave the car at the edge and continue on foot crossing the wadi which is around 700 m wide.

For more information and explanation of listwanite see EP 39.

EP 39

Listwanite contact
Topic: Carbonatised ultramafic rock, listwanite
Location: UTM 40 Q 619444 2592866 / N 23°26'28" E 58°10'09"

Rating: ☺☺

For location map see EP 37, Fig. 109.

Approach: Leave the main road #15 from Muscat to Nizwa about 5 km after the exit to Murayrat at the little village Al Batha'a. Take the road towards the south which becomes unpaved after the village. Avoid the extremely steep way uphill (pipeline) and rather follow the way around the ridge. After about 2 km you will reach the outcrop directly by the road.

Listwanite is an unusual rock type generated from ultramafic rocks by complete carbonatisation. The serpentinisation process leads to the carbonatisation of peridotitic rocks. Normally, peridotites are not completely carbonated but contain magnesite and carbonate, partly also dolomite veins. These veins can often be observed in outcrops of altered peridotites as decimeter thick white veins in the dark host rock (cf. EP 01). In case of the interaction of CO_2-rich fluids with peridotites at temperatures up to 200 °C it may occur that all magnesium and calcium as well as some of the iron reacted with CO_2 to form secondary carbonate minerals such as magnesite and calcite. An intense red colour is generated by the formation of iron oxide minerals. Listwanite is mainly composed of carbonate minerals, quartz which is derived from the altered silica minerals and sometimes also talc and Cr-muscovite. In the field, listwanite is characterised by intensely red coloured rocks on top of a peridotitic sequence, often related to faults and fractures since these structures permit the percolation of the CO_2-rich fluids through the peridotite.

EP 40

Nummulitic limestone
Topic: Eocene Seeb formation rich in nummulites overlain by Quaternary deposits
Location: UTM 40 Q 617735 2608285 / N 23°34'50" E 58°09'13"

Rating: ☺☺

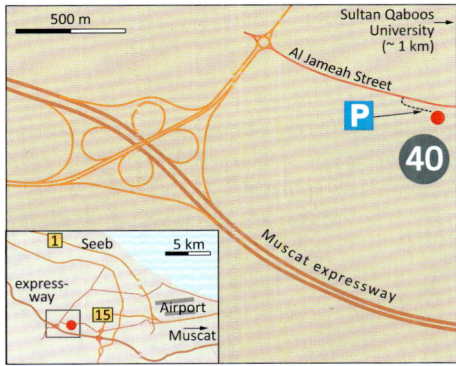

Fig. 113: Location map.

Approach: The outcrop is located in Al Khuwd close to Sultan Qaboos University. Take the Muscat expressway and leave the highway at the exit "Al Khuwd/Al Seeb". Follow the signs to "Sultan Qaboos University". Turn right after 300 m at the first roundabout and follow the road for 1 km. You will find the outcrop on the right hand side of the road on a parallel dirt road.

The outcrop is a road cut and continues for several hundred meters. Two different lithologies can be observed: a yellow highly bioclastic nodular marlstone at the base, un-

Fig. 114: Seeb formation rich in nummulites of the Eocene overlain by Quaternary deposits.

Fig. 115: Collection of nummulites from the Eocene Seeb formation.

conformably overlain by a grey conglomeratic unit. The marlstone is of middle Eocene age (Middle Lutetian to Bartonian) and is mapped as the Seeb formation (Beavington-Penney et al. 2006). The road-cuts bordering the Muscat Expressway to the east are made up of the same formation. However, they are not so rich in fossils as in this outcrop. The conglomeratic unit overlying the Eocene marlstone is an alluvial fan deposit and is indicative of recent land movement as the deposits are terraced (Fig. 114). They are probably of Quaternary age; so far the absolute age is not confirmed by dating.

The most interesting observation at this outcrop is the occurrence of abundant foraminifera (Fig. 115). They can be found in millions within the Eocene marlstone as well as weathered in the talus deposits. Foraminifera are fossilised marine organisms. They are single-celled and appear as lentil or disk-shaped within the outcrop. The largest ones are flattened discs and reach up to 6 cm in diameter. These animals have a complex internal structure and belong to the genus *Nummulites*. This term is derived from Latin, where "nummulus" means "small coin". Given the size of the *Nummulites* these would be rather large coins indeed.

EP 41

Barchans
Topic: Desert formation, dunes, barchans
Location: UTM 40 Q 764780 2422687 / N 21°53'18" E 59°33'45"

Rating: ☺

Approach: The barchan dunes are located along the road #35 close to Al Ashkharah. A number of dunes are well exposed, mainly on the western side of the road about 2 km north of the village of Al Ashkharah. An unpaved road leads to the largest dune.

Barchan dunes are crescent-shaped isolated sand dunes that develop in areas where sand is sparse and the wind is unidirectional. For different types of dune morphology compare EP 32 (linear dunes) and EP 47 (star dune). Barchans are highly mobile and their longest axis is orientated perpendicular to the wind direction. They are characterised by a large central part and two horns on the side in plan-view (see Fig. 116). These horns point downwind. Barchans are asymmetrical with a gently inclined windward slope and a steeper lee slope or slip face. The general movement of the dune is caused by saltating sand grains that are entrained on the windward side and accumulate behind the crest. Once a certain threshold value is exceeded, a sheet of sand grains avalanches down the slip face. Internally these sand grains form cross beds. The angle of response is approximately 30–35 degrees for medium-fine dry sand. As the central part is composed of a larger volume of sand, it moves slightly slower than the sides, hence the lateral horns develop.

The dune at this stop is approximately 20 m high and is moving gradually towards the NNE. Comparison of satellite images covering the period from 2004 to 2013 led to the

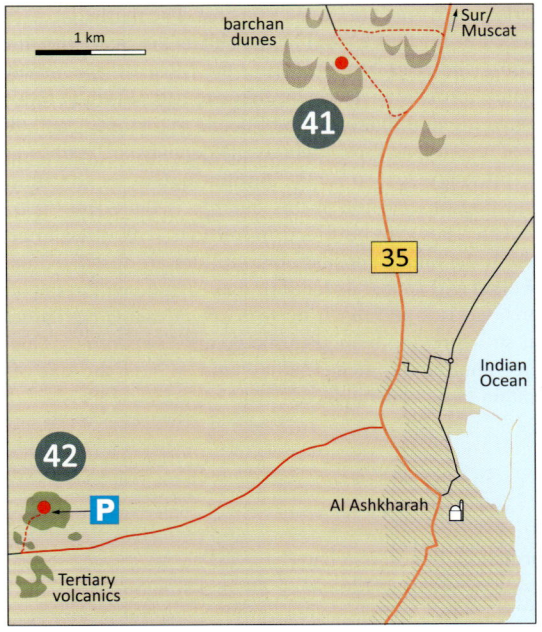

Fig. 116: Location map

conclusion, that the dune is moving with a speed of approximately 10 m per year. Neighboring dunes move with a similar speed, the fastest one with approximately 15 m per year.

EP 42

Al Ashkharah Volcanoes
Topics: Tertiary volcanic rocks, xenoliths
Location: UTM 40 Q 761776 2418044 / N 21°50'48" E 59°31'57"

Rating: ☺☺

For location map see EP 41, Fig. 116.

Approach: Follow road #35 from Al Kamil Wal Wafi towards Al Ashkharah. In Al Ashkharah turn right at UTM 40 Q 764974 2419004 / N 21°51'18" E 59°33'50" and follow the dirt road for about 3.8 km. Turn right and approach the outcrop about 400 m towards the north. Around this point and beyond, you will be surrounded by a number of Tertiary volcanoes.

Extrusive igneous rocks from the Paleogene (or Lower Tertiary) occur in different places in Oman, including Muscat and Ad Duqm. The best place to see these rocks is around Al Ashkharah area in the Wilayat of Jalan Bani Bu Ali, where many ancient volcanos exist. They occur along two main fault directions, trending approximately North-South and East-West. The timing, structural position and the source region of the Omani Tertiary basanites which are similar to basalt, but with low silica and high percentages of alkaline elements, suggest that they were the result of melting asthenospheric mantle in response to local Cenozoic lithospheric extension predating the Red Sea rift. This was probably related to the obduction of the Masirah Ophiolite and the evolution and opening of the Gulf of Aden during the Cenozoic.

The alkaline lavas from the Al Ashkharah area, facing the Indian Ocean along the North-East Oman coastline, contain numerous small (<2 cm) xenoliths, which represent rock fragments from the mantle that were incorporated into the magma when it was still molten. The mantle xenoliths provide a unique opportunity to understand the nature and evolution of the upper mantle (lithosphere) beneath the Oman passive margin, bordering the Owen Mid-Oceanic Zone in the Indian Ocean (Grégoire et al. 2009). They show strong evidence for pervasive metasomatism, perhaps in association with Paleogene tectonic and igneous activity (Nasir et al. 2006). The data obtained from analysing the mantle xenoliths suggest that eastern Arabian lower crust is of hotspot origin, in contrast to western Arabian lower crust, which mostly formed at a convergent plate margin (Nasir et al. 2006). The petrogenesis of the Omani Tertiary basanites is explained by partial melting of an asthenospheric mantle protolith during an extension phase predating the opening of the Gulf of Aden and plume-related alkaline volcanism. The mantle xenoliths are mainly spinel dunite, spinel lherzolite and spinel wehrlite.

The Omani mantle xenoliths seem different in type and chemistry and oxidation state to the related Yemeni, Saudi Arabian, Syrian, and Jordanian mantle xenoliths. The present suite of mantle xenoliths from Oman indicate that the mantle beneath Oman is homogeneous on the small scale and moderately oxidised when compared with the xenolith suite from other parts of the Arabian plate. Xenolith data and geophysical studies indicate that the Moho, which separates the crust from the mantle, is located at a depth of ~40 km.

EP 43

Aeolianite
Note: A 4-wheel vehicle is required for this excursion.
Topic: Desert formation, aeolianites, Quaternary climate change
Location: UTM 40 Q 703310 2341913 / N 21°10'02" E 58°57'29'"

Rating: ☺☺

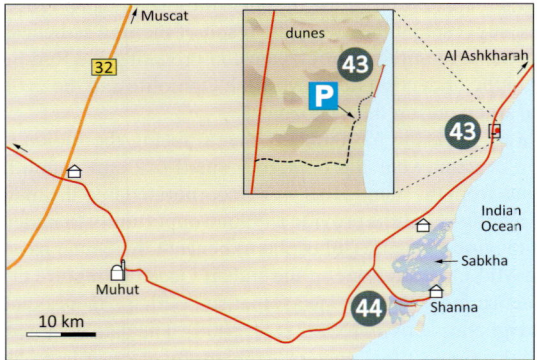

Fig. 117: Location map

Approach: Follow the coastal road from Al Ashkharah along the coast line of the Indian Ocean toward the south-west for about 100 km. Leave the road at UTM 40 Q 702783 2342663 / N 21°10'25" E 58°57'10" and drive through the sand dunes to the coast (~ 1 km). Be aware of getting stuck in the sand! The outcrop is directly at the beach. The best time to access the outcrop is during low tide.

Fig. 118: Aeolianites with cross-bedding at the coastline south of Al Ashkharah.

Exposed along the beach section are aeolianites, weakly cemented dune deposits which according to Garnder (1988) underlie the entire Wahiba sands (see EP 32). The cliff reaches more than 10 m in height locally. The most striking features are the bedding structures which can best be observed in the early morning hours when they are nicely illuminated by the rising sun (Fig. 118). Large scale cross bedding structures with decimeter to meter thick tabular planar cross beds made up of curved laminae represent the slip face of the former dune (cf. EP 41). Climbing ripple lamination as well as trough cross-beds are also common.

EP 44

Sabkha
Topic: Salt flats, evaporite formation
Location: UTM 40 Q 673395 2294514 / N 20°44'32" E 58°39'55"

Rating: ☺

For location map see EP 43, Fig. 117

Approach: Follow the coastal road from Al Ashkharah along the coast line of the Indian Ocean toward the south-west for about 110 km. Turn left at UTM 40 Q 669769 2298579 / N 20°46'45" E 58°37'51" and follow the road 5 to 6 km into the sabkha landscape.

The road to Shenna, where the ferries for Masirah Island leave, leads through a landscape barren of vegetation that shows little topography with some ponded depressions. The only elevations in this extremely flat terrain are sandbars close to the open sea. The Arabic term "sabkha" refers to salt flats (cf. Al-Farraj 2005). They form along arid coastlines as a result of relative sea level drop and occasional flooding. It is a supra-tidal environment that usually develops out of a lagoon, locally known as Khor. Marine lagoons commonly border on the seaward side of the sabkhas. The dominating sediments are evaporates such as salt (Fig. 119), gypsum and carbonates with intercalated aeolian siliciclastics. Algal-microbial mats that resemble dried rags can be observed at the shorelines of the salt flats. Trenching in to the sabkha reveals lamination which is mainly caused by alternating algal mats and clastic deposits (Fig. 120). The organic material of the algae may be considered as a possible source of hydrocarbon reservoirs (Homewood et al. 2007). Therefore, these recent systems are used as modern analogs for older formations important in the hydrocarbon industry. There is ongoing salt farming along the sabkha (Fig. 121).

Fig. 119: Sabkha formation with salt evaporites. The inset shows a dried algal mat at the shoreline of the abkha.

Fig. 120: Shallow pit (scale-bar is 25 cm) reveals the internal stratification of the sabkha, characterised by bright layers of siliciclastic layers and dark layers of algal-microbial mats.

Fig. 121: Salt mining activities within the coastal sabkha. The salt is scratched from the surface and piled up to mounds of 50 cm height.

EP 45

Miqrat sandstone
Note: You will need a 4-wheel vehicle for this excursion. It is also advisable to visit the site with more than one vehicle, especially if the area has experienced recent rain.
Topic: Colourful sandstone deposits
Location: UTM 40 Q 604707 2323592 / N 21°00'35" E 58°00'27"

Rating: ☺

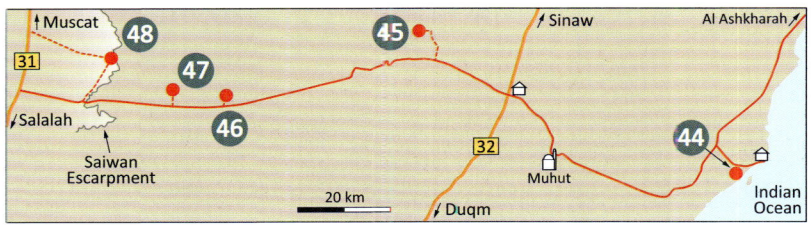

Fig. 122: Location map.

154 Field sites

Approach: Follow the main road #32 from Sinaw to Duqm for about 190 km up to a junction with an east to west trending road (at UTM 40 Q 624123 2309248 / N 20°52'45" E 58°11'36"). Turn right and follow the paved road for another 18 km to UTM 40 Q 608016 2316822 / N 20°56'55" E 58°02'20". Turn right onto an unpaved road and stay on this road for 7 km. To reach the outcrop you have to leave the road and drive west for about 2.5 km. The outcrop is well visible in the desert.

With the onset of the Palaeozoic period, a succession of siliciclastic and carbonate rocks up to several kilometers thick were deposited across Oman following a major tectonic phase related to the collision of the eastern and western parts of a supercontinent known as Gondwana that existed around 500 million years ago. During the collision, Oman was uplifted above sea level and a sequence of mainly siliciclastic rocks, known as the Haima Supergroup (Cambrian-Silurian), was deposited. The main clastic source areas were in the south in a major landmass of Gondwana and the depositional setting was initially continental with a more marine setting developed higher in the sequence.

Fig. 123: Colourful sandstones of the Miqrat formation.

The outcrops of the Haima Supergroup form colourful laminated ridges in the northern part of Huqf that get eroded and weathered into different shapes. Among the most famous sculptures formed naturally in the area is the Omani sphinx and pyramids that are formed from the outcrops of the Miqrat formation (Fig. 123). This formation is dominated by sabkha. Sabkha is the Arabic term for a salt flat or playa (see EP 44). The Miqrat formation is mainly made up of thick packages of red mudstone and siltstones that are separated by very fine-grained sandstone. The different colours of the layers represent an interplay between wet and dry climate cycles that influence the water table level and the sedimentation rate. Overall, within the Lower Palaeozoic sequence of interest in northern Oman the complete spectrum of alluvial (fluvial, sheet-flood, aeolian) hydrocarbon re-servoirs are likely to be developed within the Miqrat formation in the sub-surface (Millson et al. 1996). Therefore, the outcrops of Miqrat in this area are important as subsurface analogues for petroleum geologists and engineers. The interesting thing is that the sabkha and sand dune environment that resulted in the deposition of the Miqrat formation about 500 million years ago can be seen and examined around the outcrops of this formation today. You might want to explore the similarity and the cyclicity of the present system by digging a trench in the surrounding present sabkha (cf. EP 44).

EP 46

Permian Khuff formation
Topic: Clastic sediments of Carboniferous to Permian
Location: UTM 40 Q 561576 2307084 / N 20°52'46" E 57°35'30"

Rating: ☺

For location map see EP 45, Fig 122.

Approach: Follow the main road #32 from Sinaw to Duqm for about 190 km up to a junction with an east to west trending road (at UTM 40 Q 624123 2309248 / N 20°52'45" E 58°11'36"). Turn right and follow the paved road for another 65 km. Turn right into an unpaved road after crossing a hill with a nice road cut. The outcrops are on the flat hills on both sides of the dirt road about 500 m north of the main road.

The stop is located in the middle of the desert landscape of Huqf. The outcropping silicified limestone and cross-bedded sandstone are part of the Carboniferous – Early Permian Khuff formation which is part of the Akhdar group. This formation is dominated by clastic sediments representing a shelf facies. Pieces of silicified wood which are mainly parts of tree trunks can be found scattered on the surface. Some layers show spectacular hardground horizons with abundant fossils such as bryozoans, bivalves, gastropods, nautiloids (*Orthoceras*) and crinoids (Fig. 124). In some cases, the orientation of the fossil remains indicates the direction of palaeo-currents. The surfaces of the rocks show desert varnish and polish which makes the fossils aesthetically particularly

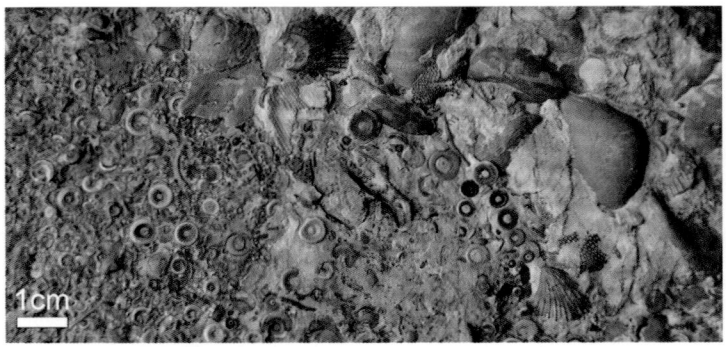

Fig. 124: Fossils of the Early Permian Khuff formation on a hardground horizon.

attractive. Some horizons are densely packed with the mm-sized spiriferinide brachiopod *Pachycyrtella omanensi*, a stationary suspension feeder adapted to sandy substrate with high energy flow. Furthermore, nodular sandstone with spherical objects of 1–2 cm diameter can be observed. The structures might be explained by bioturbation or as a diagenetic effect, but this is still hypothetical at the moment.

EP 47

Star dune of Saiwan
Topic: Dune formation and morphology
Location: UTM 40 Q 550607 2309588 / N 20°53'08" E 57°29'11"

Rating: ☺☺

For location map see EP 45, Fig. 122.

Approach: Continue on the paved road from EP 45 towards the west for another 12 km. Turn right (at UTM 40 Q 550047 2306698 / N 20°51'35" E 57°28'51") and drive for about 3 km on the uncovered desert floor towards the north. The best approach to the dune is from the north. You need about 10 to 15 minutes to climb up to the top.

The star dune of Saiwan is a prominent landmark and part of several isolated dune forms that are located at the margin of a larger dune field made up of linear dunes to the north (Fig. 125). The dune is composed of quartz rich yellow sand and has a maximum height of around 70 m whereas the body of the main dune is 500 m across. Star dunes form due to changing wind patterns. These dunes are pyramidal sand mounds with slip

Fig. 125: Star dune of Saiwan, view towards south.

faces on various sides radiating away from the central part, indicating a multi-directional wind regime. Therefore, in contrast to barchans (see EP 41) star dunes do not migrate.

EP 48

Lake Saiwan
Topic: Escarpment, Palaeolithic site
Location: UTM 40 Q 537433 2317395 / N 20°57'24" E 57°21'36"

Rating: ☺☺

For location map see EP 45, Fig 122.

Approach: Continue on the paved road from EP 47 towards the west for another 19 km. Turn right (at UTM 40 Q 531363 2307013 / N 20°51'46" E 57°18'06"). There are numerous tracks crossing the deflated surface which is easy to travel on. Deflation here refers to erosion that occurs by the force of the wind removing loose and fine-grained particles from the surface of the earth. Drive towards the coordinates of the stop and try to avoid the escarpment to your right.

The stop is on top of a 20 to 30 m high steep limestone escarpment (Fig. 126) with Early Miocene Lower Fars Group in shelf, fore-reef and slope facies at the base, overlain by the Middle Miocene Fars Group in lacustrine facies. The limestone layers are almost perfectly horizontal, resulting in parallel escarpment retreat.

Relict lake deposits of Pleistocene age crop out in the depression in front of the escarpment. Inhomogeneous conglomerates and breccias as well as fluvial deposition with aeolian intercalations indicate a lake shore environment with oscillating water tables. The palaeo-lake had a size of roughly 1400 km^2 with a maximum water depth of 25 m. Optically stimulated luminescence (OSL) ages of the lake deposits attest a depo-

Fig. 126: The Saiwan Escarpment in Middle Miocene limestone.

sitional age of 120+/−8 ka confirming the existence of a Pleistocene wet phase (Rosenberg et al. 2012). These palaeo-lake deposits indicate favorable environmental conditions very different from the present hyper-arid environment.

Archaeological evidence such as flint tools and workshops indicates substantial, dense and lasting past human occupation along the lake shores (Biagi 1994b). It is speculated that the remains are from an independent Late Pleistocene culture in southern Arabia as similar tools are unknown from other areas. It remains speculative at the moment whether or not the group was an endemic population or related to the anatomically modern humans. The latter would support a southern out-of-Africa migration route (Fig. 2).

EP 49

Qarat Kibrit
Salt diapir, Neoproterozoic (Ediacara) and Cambrian sediments, oil reserves
Location: UTM 40 Q 515447 2378680 / N 21°30'27" E 57°08'56"

Rating: ☺☺

Approach: Follow the Salalah-Muscat highway (roads #15 and #31) to Adam, about 200 km southwest of Muscat and continue on road #31 to the south, entering the desert. You will reach the Ghaba resthouse 105 km south of Adam (UTM 40 Q 515447

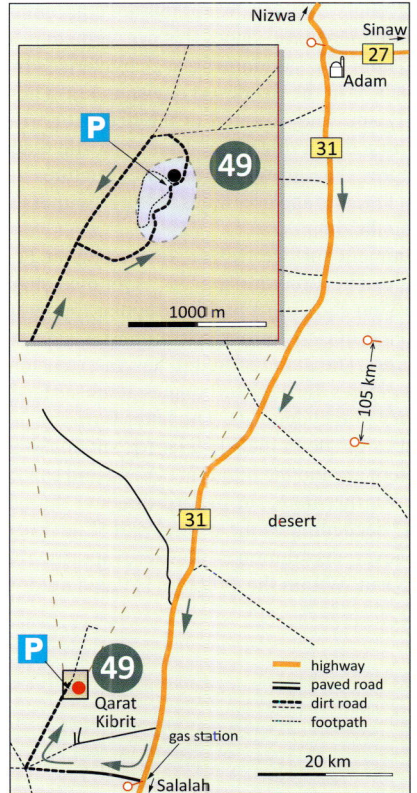

Fig. 127: Location map.

2378680 / N 21°22'39" E 57°15'20"). Turn to the west onto a dirt road and after 15 km turn right to the north-northeast. Follow the road for another 14 km until you reach the salt dome located on the right hand side of the road. You can enter the center of the dome by 4-wheel car only.

There are six surface piercing salt domes in Oman. These salt domes or diapirs form when salt rock intrudes vertically in overlying denser rocks. Here, they derive from the Precambrian Ara Group salt formation (Peters et al. 2003). Sedimentation began in the late Proterozoic and lasted until the Early Cambrian in the northeast of Gondwana. The hills formed by the salt diapirs have elevations of 100 m or less above the surrounding areas and are up to 3 km in diameter. They are situated within an area of Quaternary gravel desert which is strongly deflated. Ventifacts are common features on the surface. Ventifacts are stones with interesting shapes that are formed by sand which is blown in the wind and causes erosion. The salt domes are characterised by a rugged morphology and stand out as isolated hills in the flat desert environment. These diapirs are structures within the Ghaba Salt Basin. They formed in the deepest part of the basin where the base is estimated to be in a depth of 10 km. Therefore, the diapirs are regarded as extremely high-relief features. The growth of the salt structures started in the Early Palaeozoic along deep-rooted faults and continues today as witnessed by Quaternary drainage patterns surrounding the diapirs.

Qarat Kibrit is a prominent surface feature, located about 11 km west-northwest of the Ghaba North oil field. The structure appears as an oval shaped form with an outer rim and a central depression showing a highly irregular topography (Fig. 128). The topography is further altered due to former salt-mining. Caves within the structure are evidence of these activities. The diapir structure has an extension of 700 × 300 m and is 30 m high. Qarat Kibrit pierces Tertiary carbonates at the surface. These are overlain by a thin layer of Quaternary gravel deposits. The latter form the distal part of a large system of alluvial fans that originate in the Al Hajar Mountains to the north.

A radial drainage pattern around the diapir is evident and can be seen clearly on satellite images (Fig. 129). Furthermore, an inward-directed drainage system developed, which contributes to the deepening of the central part (Fig. 130) of the dome-structure by salt dissolution. However, dissolution is slow as evaporitic rocks are exposed. Nowadays, the area is hyper-arid. Within the structure isolated outcrops of mainly salt, gyp-

Fig. 128: Overview of the Qarat Kibrit salt dome structure. View from the western rim towards the southeast. Fine laminated black carbonates from the Ara Group can be seen in the foreground.

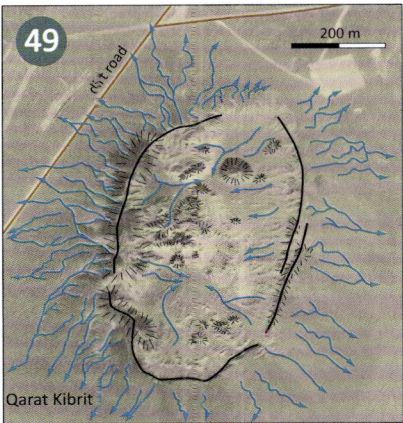

Fig. 129: Radial drainage pattern around the diapiric structure of Qarat Kibrit.

sum and anhydrite occur. Beside these evaporitic rocks, carbonate rocks are exposed as exotic blocks. They represent intra-salt "stringers" and formed as isolated carbonate platforms in the deeper parts of the basin (Reuning et al. 2009, Schoenherr et al. 2010). The sediments within the Ara Basin were initially deposited horizontally. However, due to the diapirism their arrangement is more or less chaotic in the surface outcrops.

The rocks of the Ediacaran – Early Cambrian Ara Group were deposited in six full cycles of salt and carbonates until they were overlain by Cambrian sediments. Different facies types are recognised within the outcrop. They compromise different evaporitic rocks as well as finely laminated dark carbonates (Fig. 131): thrombolites, grainstones, organic-rich laminated dolostones and laminated stromatolites representing the stringer carbonates. Oil source rocks, mainly mudstones, formed in the deeper parts of the basin.

Fig. 130: Central part of Qarat Kibrit.

Fig. 131: Contact between Precambrian salt and fine laminated black carbonates from the Ara Group in the center of the salt dome structure of Qarat Kibrit.

Worldwide, salt-diapir flanks are well-known for hydrocarbon provinces. However, this is not the case with the surface piercing diapirs in the Ghaba Salt Basin. Substantial reserves have so far only been discovered in structural culminations overlying the crest of buried salt swells and diapirs.

EP 50

Tower tombs and travertine at Halban
Topic: Archaeological site – tower tombs and travertine formation
Location: UTM 40 Q 606073 2608014 / N 23°34'44" E 58°02'22"

Rating: ☺

Approach: Leave the Muscat expressway at the exit to Halban and follow the newly constructed road for about 5 km to Halban. Take the road to the east at the roundabout and follow this road for 1.7 km to its end. Continue straight ahead into an unpaved road for 200 m. The tower tombs are a walking distance of 100 m away.

Field sites

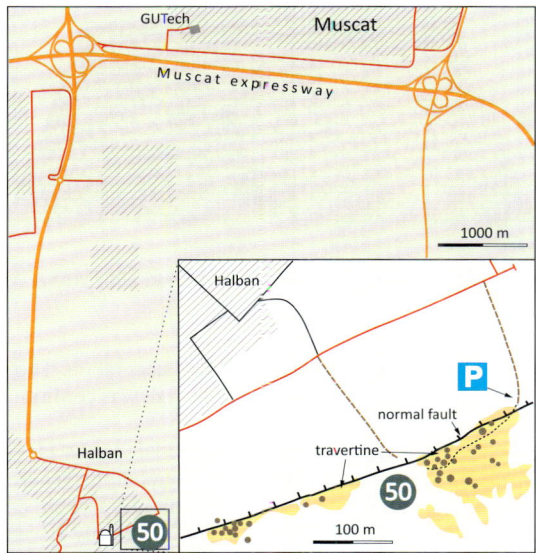

Fig. 132: Location map. The round dots in the inset map represent the locations of the tombs.

The tower tombs, also called cairn (stone pile) graves, next to Halban are dated back to the Early Bronze Age (~3000 BCE). Another name sometimes used is "beehive tombs" which is derived from their beehive-like appearance (compare Fig. 63 of EP 13). At Halban, more than 30 tombs are located on a travertine plateau (Fig. 132) overlooking the village and the Batinah plain. This type of tomb is usually dated to the so-called Hafit period, around 3,300–2,700 BCE. Their sizes range between 4 and 9 meters in diameter, whereas the internal funerary chamber does not exceed 2 meters. The tombs narrow toward the flattened top and appear – if they are well preserved – in a conical shape, reaching 4 to 8 meters in height (Fig. 133). The rectangular to trapezoidal entrance face to the east in most cases. Building materials are either worked stone slabs or undressed stones that were piled up to two to three walls. The stones were assembled without using mortar. In Halban, the builders used carefully dressed local travertine and constructed one inner and a second facade wall. The tombs at Halban are heavily destructed; the tomb entrance is buried nowadays as the eastern side is the most severely damaged part.

Hafit tombs functioned as places for collective burials of families and were probably used over generations. Although most of the cairns were plundered over time, grave artifacts like beads, copper rings and pottery were discovered in some tombs and provide at least a glimpse of the rituals and beliefs of Bronze Age societies. The grave artifacts also allowed dating of the tombs. Pots, which occasionally were among the

Fig. 133: Sketch of a reconstructed tower tomb (modified after Yule & Weisgerber 1998).

items, were identified by their fabric, shape, size and decoration. They are identical to those made in Mesopotamia around 3,000 BCE. This age was confirmed by carbon dating.

The travertine, which was used as building material, forms elevated terraces on top of ultramafic rocks mapped here as tectonised harzburgite. These mantle derived rocks belong to the ophiolite nappe of the Samail Ophiolite. The travertine indicates an ENE-WSW striking normal fault. The fault dips by 60° steeply toward the northwest. (Fig. 132) and can be traced over several kilometers along the rim of the dark ophiolitic mountains. The occurrence of travertine at this location is bound to the fault and is a result of circulating calcareous water moving along the fault zone.

EP 51

Nakhl Hot Spring (Ayn A' Thowarah)
Topic: Thermal Spring
Location: UTM 40 Q N 584666 2585327 / N 23°22'31" E 57°49'43"

Rating: ☺

For location map see EP 58, Fig. 145.

Approach: Leave the main road #13 in Nakhl at the central roundabout towards the east, following the road for 4 km through the city of Nakhl to the hot springs which is signposted.

The springs in Nakhl were mentioned by early explorers as a remarkable site in a desert environment (see Miles 1901). The main thermal spring in Nakhl has a temperature of 38.5 °C. Its Arabic name is Ayn A' Thowarah which means "spring that boils". The spring gushes out from Early Cretaceous limestone at the southern valley side of a wadi. Dobretsov et al. (2011) measured a pH value of 7.9 and a flow velocity of 2.4 m/s.

Ayn A' Thowarah is one of several springs which are located north of the Al Hajar Mountains, all associated to a fault line at the foot of the mountains. The spring in Nakhl is one of the most productive. The place is a popular picnic spot and becomes crowded, especially on weekends. The water is collected in an artificial pool and distributed to the Falaj "Kabbah" and Falaj "As Sarooj". It is well worth exploring the oasis itself with its lush greenery of palm-, banana-, mango-, and pomegranate trees, the water supplying falaj and ruins of old mud-brick buildings scattered around. This can be done comfortably on foot.

EP 52

Wadi Mistal – landslide
Note: The parking site inside the wadi might be not accessible due to recent changes in the road layout. It is also possible to leave the car on the main road at the entrance to the wadi and walk down.
Topic: Landslide
Location: UTM 40 Q 570954 2578444 / N 23°18'50" E 57°41'38"

Rating: ☺☺

Approach: Follow the road #13 from Nizwa towards Rustaq. Take the turn-off towards Wadi Mistal which is signposted; also "Wadi Mistul". You will approach the mountains and arrive at the narrow gorge that marks the entrance into the wadi. Leave the car at the bottom of the wadi at UTM 40 Q 571004 2578587 / N23°18'54" E 57°41'40" and continue walking into the wadi bed.

Wadi Mistal is one of the major valleys that cut through the northern flank of the Al Hajar Mountains. The section starts with mainly carbonatic Cretaceous deposits at the entrance to the wadi. Structures such as as tension gashes in decimeter-wide shear zones or thrust faults indicating several meters of displacement (Fig. 135) are well exposed in the high and steep walls of the wadi. Furthermore, fossils of rudists, also known as marine heterodont bivalves, can be observed within the thick banked limestone. The rock units get successively older as one moves further into the valley. Permian rocks prevail before the narrow valley opens into the large Gubrah Bowl (see EP 54). This large depression is located in the center of the mountains and is 15 by 13 km in size. The bowl forms a giant natural amphitheater framed by the resistant Permian dolomites. Precambrian rocks make up the bottom of the bowl (see EP 55).

Wadi Mistal is the only outlet of the Gubrah Bowl with a catchment area of 210 km^2. The stops described at EP 52 and 53 are evidence for Pleistocene landslide damming of the wadi (see Hoffmann et al. 2015). Further stops within Wadi Mistal are described at EP 54 and 55.

The eastern side of the wadi is characterised by clastic deposits (Fig. 136). The outcrop shows two distinct units. The lower 2–3 m are dominated by (semi-) rounded

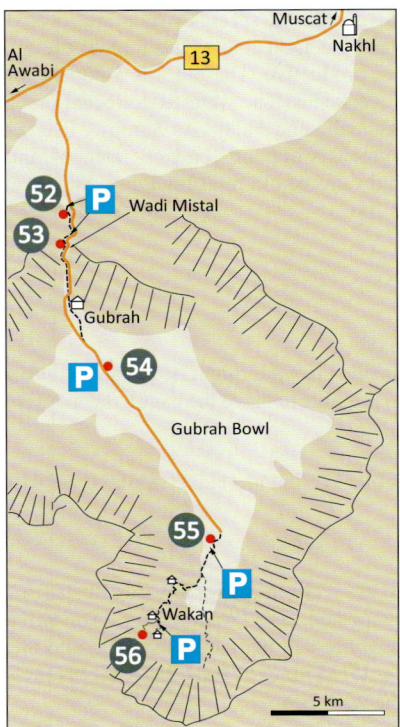

Fig. 134: Location map.

blocks which float in a silty matrix. This unit is horizontally layered. Imbrication structures within this deposit indicate fluvial transport from south to north. This matrix-supported conglomerate is interpreted to represent debris-flow deposits.

This unit is overlain by an even coarser deposit. The largest grain size observed measures 8 m in diameter. The blocks within this unit are angular; the sorting is poor and the deposit is chaotic and poorly stratified. Both units are cemented. The upper unit can be followed for 1 km on the eastern side of the bank. Huge boulders can be observed in the wadi bed and occasionally unsorted clastic deposits occur on the western side of the wadi as well. These deposits are interpreted to represent a rockfall and/or landslide deposit. The total mass of the rockfall/landslide deposit is estimated as 9 × 10^6 m^3. The deposits of this mass movement lock the wadi, causing the temporary existence of a lake the sediments of which can be studied at EP 53.

Field sites 167

Fig. 135: Overthrust in Cretaceous limestone at the northern entrance into Wadi Mistal.

Fig. 136: Rockfall/landslide deposits at the mouth of Wadi Mistal.

EP 53

Wadi Mistal – lake deposits
Note: The final approach to the outcrop described might change due to recent road work in progress.
Topic: lake sediments, palaeoclimate
Location: UTM 40 Q 570824 2576958 / N 23°18'01" E 57°41'33"

Rating: ☺☺

For location map see EP 52, Fig. 134.

Approach: Follow the road #13 from Nizwa towards Rustaq. Take the turn-off towards Wadi Mistal which is signposted, also "Wadi Mistul". You will approach the mountains and arrive at the narrow gorge that marks the entrance into the wadi. Follow the main road for 1.5 km and turn right into the wadi bed. Follow the graded road in the wadi for another 750 m. The outcrop is located on the right hand side.

The sediments in this outcrop represent erosional remnants of a more widespread deposit which probably covered the entire wadi. Several smaller outcrops showing similar deposits can be observed along the main road. The location of the outcrop is behind a bend in the course of the wadi and the deposits are therefore protected from erosion. The outcrop shows 112 fining-upward sequences in a 22 m profile, each around 5–20 cm in thickness (Fig. 137). The general grain size is fine-sand to silt. The sediments

Fig. 137: Fining upward sequences in Quaternary lake sediments of Wadi Mistal.

Fig. 138: General outcrop overview. The numbers in the lower sketch indicate depositional ages in kilo years.

are horizontally bedded; channel fills and small scale cross bedding can be observed locally. The deposits are located in a small side-valley. Further up this side-valley, the fine-grained deposits show a lateral facies transition to coarser deposits, from sand to gravel. Within the profile, distinct layers characterised by desiccation cracks can be observed that indicate temporary dry conditions. Furthermore, in-situ plant remains are identifiable which are also interpreted as representing palaeo-surfaces. The sediment sequence is dated by optically stimulated luminescence (OSL) providing a time of deposition of approximately 50,000 to 40,000 years ago. See Fig. 138 for the ages and Hoffmann et al. (2015) for more detailed information.

EP 54

Gubrah bowl – overview
Topic: Basement cored anticline, geological window
Location: UTM 40 Q 572422 2571633 / N 23°15'08" E 57°42'28"

Rating: ☺☺

For location map see EP 52, Fig. 134.

Approach: Follow the road #13 from Nizwa towards Rustaq. Take the turn-off towards Wadi Mistal which is signposted, also "Wadi Mistul". You will approach the mountains

Fig. 139: View towards the north from the Gubrah Bowl into Wadi Mistal on the left hand side of the picture.

and arrive at the narrow gorge that marks the entrance into the wadi. Follow the main road for 15 km. You will pass EP 52 and 53 on the way.

This stop is an overview of the Gubrah Bowl which represents a basement cored anticline and appears as a huge natural amphitheater (Fig. 139). The surrounding mountains have elevations of 2,000 to 2,500 m and are made up of weathering resistant Triassic to Jurassic dolo- and limestone. The center of the depression is dominated by older, Precambrian rocks. They are mainly composed of siltstone and are much less resistant to weathering. The unconformity between Permian and pre-Permian rocks (see EP 56) can clearly be identified from this stop as a change in slope in the surrounding mountains. The uplift of the mountains started in the Neogene and continues until today as evidenced by fluvial terraces within the bowl as well as in the alluvial fans north and south of the mountains.

Drastic climatic changes are encountered throughout the Quaternary, shifting from wet monsoon climates to arid and even hyper-arid conditions (see EP 52 and 53). As a result, the rocks weathered and were transported by fluvial processes to the north where they build up the Batinah coastal plain. The allochthonous nappe units (Hawasina and Samail nappe) once covered the Mesozoic sequence of the Al Hajar Mountains. They are entirely eroded in the center of the mountain chain. The fluvial gravel that dominates the surface here gives clues to this changeful erosion history. Microkarst structures such as grooves and scalloped textures may be observed on the limestone pebbles.

EP 55

Gubrah bowl – Precambrian diamictite
Topic: Neoproterozoic glaciation, snowball earth, tillites
Location: UTM 40 Q 577633 2563181 / N 23°10'32" E 57°45'30"

Rating: ☺☺

For location map see EP 52, Fig. 134.

Approach: Follow the road #13 from Nizwa towards Rustaq. Take the turn-off towards Wadi Mistal which is signposted; also "Wadi Mistul". You will approach the mountains and arrive at the narrow gorge that marks the entrance into the wadi. Follow the main road for 19 km. You will pass EP 52–54 on the way. Pass the hospital and school at the end of the Gubhra bowl. The outcrop is located on the right hand side of the recent wadi bed. Road construction was in progress at the time of writing. The outcrop is located around hundred meters north-west of the new road.

The surficial sediments within the Gubrah bowl are mainly Quaternary gravel deposits. Scattered outcrops of the Precambrian Mistal formation are located within wadis that cut into the Precambrian underground. The Mistal formation is the lowest sedimentary

Fig. 140: Diamictite, probably formed as tillite, of the Precambrian Mistal formation.

Fig. 141: Component with striations in the diamictite/tillite of the Mistal formation, Precambrian.

unit within the Jebel Akhdar (Rabu et al. 1986). Poorly stratified coarse sandstone, greywacke and diamictite conglomerate are exposed (Figs. 140 and 141). The clast sizes vary and may reach up to 1 m. The rocks are polymict and contain porphyritic leucogranite, biotite granite, rhyolite, siltstone, feldspathic wacke and algal-laminated limestone. The matrix is coarse grained. The rocks are metamorphosed and as an expression of schistosity, a slate-pencil fabric can be observed. Some of the larger clasts have parallel striations on the surface (Fig. 141). The deposits are interpreted as glaciogenic diamictites and are, therefore, also referred to as tillites (see EP 62 for a full discussion and Le Guerroué et al. 2005).

EP 56

Wakan falaj system
Topic: Agriculture, irrigation system, pillow basalts
Location: UTM 40 Q 575303 2559551 / N 23°08'35" E 57°44'08"

Rating: ☺☺

For location map see EP 52, Fig. 134.

Approach: Follow the road #13 from Nizwa towards Rustaq. Take the turn-off towards Wadi Mistal which is signposted; also "Wadi Mistul". Follow the main road into the wadi for about 20 km. You will pass EP 52 through 55 on the way. The new asphalt road ends at a village at the southern end of the Gubrah bowl. Continue on the gravel road uphill to Wakan. Leave the car outside the village at a car park and climb up the mountain on a paved footpath. At the end you will reach the source of the falaj system. To reach the pillow basalts: enter the village from the parking lot and follow the path outside the fence downhill. You will find the pillows at UTM 40 Q 575195 2559384 / N 23°08'29" E 57°44'04".

There are several small villages located at the southern rim of the Gubrah bowl. Wakan is one of the most spectacular ones. The village is located on top of a ridge at an altitude of around 1,500 m. The climate is pleasant here all year round. You will reach the agricultural terraces of the village once you pass the small mosque. Various crops are grown. Beside vegetables, there are fruit trees like apricots, almonds and pomegranates. The hiking trail follows the main falaj that allows the farming. There are several water-tanks and various techniques of this ancient and highly precise irrigation system that can be seen along the way. The end of the falaj is 900 m further (at 40 Q 574691 2559199 / N 23°08'23" E 57°43'46") where the main spring for the channel system is located at an altitude of 1,600 m. The system is fed from contact springs that are situated at the unconformity between the Precambrian rocks and the overlying Permian-Mesozoic strata. The Permian-Mesozoic rocks are predominantly composed of limestone which are intensively karstified and, therefore, permeable. Travertine deposits mark the location of the springs. The Saiq Plateau is located directly south of the mountain chain. Precipitation here is about the highest in the country (ca. 350 mm/year) due to the orographic effect. Views back into the Gubrah bowl reveal the unconformity along the scarp.

Another stop of interest are pillow basalts outcropping below the village (Fig. 142). These have sizes of up to 1.5 m, are slightly tilted but not deformed and still have their convex shape. This is surprising as they are Proterozoic in age (Saqlah Member of the Abu Mahara formation, see Allen 2007). Exfoliation joints as a consequence of thermo-elastic strain are observable on the rocks' surfaces. Le Guerroué et al. (2005) describe the rocks as "…basalt composed of microlithic lava with a spilitised andesitic basalt

Fig. 142: Proterozoic pillow basalts in Wakan. Photo: courtesy of Peter Nievergelt.

composition. Vesicles <1 cm size, commonly infilled with cryptocrystalline chlorite associated with carbonate, form amygdales." The morphology of the pillows is characteristic for subaqueous eruptions and probably indicates an initial rifting phase.

EP 57

Wadi al Abyad – calcite precipitation
Topic: Alteration of ultramafic rocks, calcite precipitation
Location: UTM 40 Q 568667 2590626 / N 23°25'26" E 57°40'20"

Rating: ☺☺☺☺

For location map see EP 58, Fig. 145.

Approach: see EP 58

Wadi al Abyad means "the white wadi" and is one of the wadis in Oman that is water-bearing throughout the whole year. The wadi cuts through ophiolitic rocks yet transports mainly light greyish pebbles. Therefore, the river bed is coloured almost white in contrast to the surrounding ophiolites. Another explanation for the name of the wadi are several ponds with diameters up to several meters which are covered with a white

Fig. 143: Recent calcite precipitation in little pools in the Wadi al Abyad.

Fig. 144: Carbonate cemented river gravel overlying ultramafic rocks in the Wadi al Abyad.

precipitate that floats like ice on the water (Fig. 143). The precipitate continuously grows in thickness and, once it is too heavy, sinks to the bottom.

Hot, hyper-alkaline springs that are strongly enriched in calcium are found connected to those ponds. Here, gaseous bubbles can be observed. The water smells and tastes of H_2S. The carbonate cementation is a consequence of the alteration of olivine and pyroxene to carbonate and serpentine minerals, i.e. the process of serpentinisation. The main primary precipitates in Wadi al Abyad are aragonite and brucite (Chavagnac et al. 2013). The carbonate cementation is best seen in the thick layers of carbonate cemented river gravels in the wadi (Fig. 144). For further explanation of the carbonate cementation see EP 88.

EP 58

Wadi al Abyad – Moho
Topic: Mohorovičić discontinuity, ultramafic rocks and gabbros of the Samail Ophiolite
Location between   UTM 40 Q 567402 2593033 / N 23°26'47" E 57°39'35"
          and   UTM 40 Q 568715 2589994 / N 23°25'06" E 57°40'16"

Rating: ☺☺☺☺

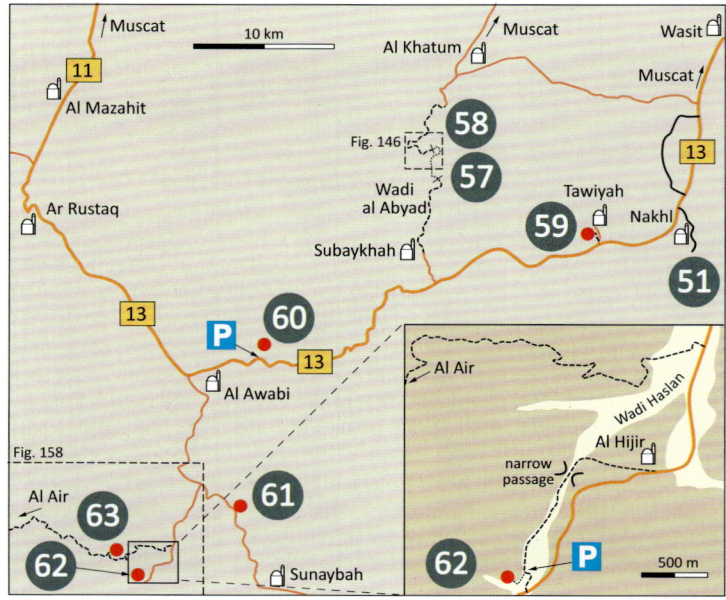

Fig. 145: Location map.

Approach from the north: (1) Leave the main road #1 (A'Seeb Street) west of Muscat at Barka towards the south-southwest to Hibra in the direction of Nakhl and follow road #13 for about 21.7 km to Hibra. Turn right towards the west and follow the Nakhl-Abyad road for about 19 km to Wadi al Abyad. Or (2) leave the main road #1 10.7 km west of the roundabout in Barka and follow the Abyad road which is not numbered, for 26 km southward to the Nakhl-Abyad road. Turn right into the Nakhl-Abyad road and follow it for 1.5 km. Both approaches now turn left into the Wadi al Abyad and use the road towards the south. It becomes unpaved after about 1.7 km and the outcrops start after another 1.5 km south and south-westward in the Wadi al Abyad. The unpaved road is drivable for roughly 4 km but it depends on the water level in the wadi. The walking distance to some spectacular outcrops of the Moho and mantle rocks is about 1.5 km (Fig. 146).

Approach from the south: Leave the main road #1 (A'Seeb Street) west of Muscat at Barka towards the south-southwest to Nakhl and follow road #13 about for 48 km passing through Nakhl. 18 km after Nakhl turn right into a newly paved road to the village Subaykhah. After Subaykhah, the road runs into the Wadi Al Abyad and becomes unpaved. Follow the drivable wadi for about 5 to 6 km towards the north. From there the walking distance is about 1 km to the southern outcrops of the Moho (Fig. 146).

A detailed overview of the most interesting points in the Wadi al Abyad is given in Fig. 146. The best outcrops of the fossil Moho and some other petrographic sites worth seeing (partly described at EP 57) are given in the following table (from north to south):

Outcrop, location	UTM coordinates	Geographic coordinates
Layered gabbro (Fig. 150)	40 Q 567058 2592951	N 23°26'42" E 57°39'23"
Ultramafic cumulates (Fig. 149)	40 Q 566778 2592695	N 23°26'34" E 57°39'13"
Campsite	40 Q 567143 2592655	N 23°26'33" E 57°39'26"
Fossil Moho (Fig. 148)	40 Q 567351 2592697	N 23°26'34" E 57°39'33"
Wehrlite outcrop	40 Q 568297 2591994	N 23°26'11" E 57°40'06"
Wehrlite intrusion in gabbro	40 Q 568224 2591992	N 23°26'11" E 57°40'04"
Terrace of calcite precipitation (EP 57)	40 Q 568245 2591938	N 23°26'09" E 57°40'04"
Fossil Moho	40 Q 568264 2591892	N 23°26'08" E 57°40'06"
Calcite precipitation (EP 57)	40 Q 568667 2590626	N 23°25'26" E 57°40'20"

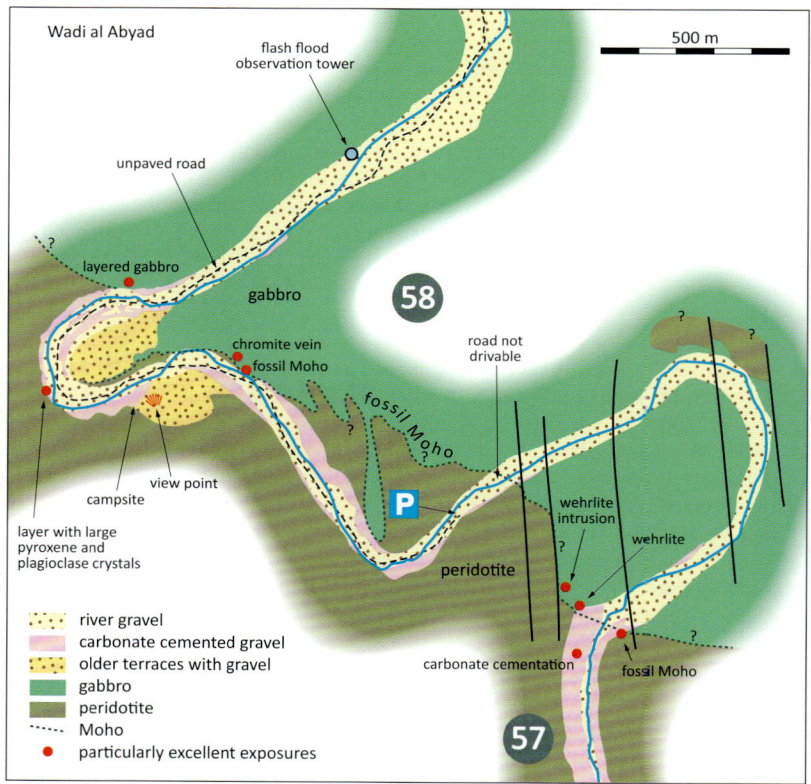

Fig. 146: Detailed map of the outcrops in the Wadi al Abyad. Geological mapping after Rabu et al. (1986), MacLeod & Yaouancq (2000), Friedrich (2015, pers. comm) and own observations. The contact between gabbros and ultramafic rocks is interpreted to represent the fossil Moho.

The outcrop of the fossil Moho is regarded to be one of the world's best (Fig. 147). Clearly distinguishable series of gabbroic and peridotitic rocks occur up the walls of the wadi. Numerous cumulate structures and dikes cutting through the primary rocks are exposed in the direct contact zone of gabbros and peridotites.

The rocks of the Samail Ophiolite formed at a mid-oceanic spreading center in the Cenomanian (95.3 ± 0.2 Ma – age data from plagiogranites as the assumed last formation event of oceanic crust; Warren et al. 2005). The spreading zone was part of the Neotethys Ocean which was converted into a subduction zone in the Middle Cretaceous (Fig. 17). The ophiolite is mostly interpreted as having originated at a back arc spread-

Fig. 147: Classical outcrop of the fossil Moho in the Wadi al Abyad.

ing center in a supra-subduction zone (SSZ) setting which is accompanied by a high temperature metamorphic sole (Glennie 2005). On the other hand, arc-related volcanic rocks have never been described in the Hawasina basin that has been overridden by the ophiolite nappe. Therefore, it seems difficult to maintain the interpretation as a SSZ origin of the spreading ridge and it may be advisable to revert back to the original definition of obduction by Coleman (1971, Nicolas 2016).

In the Wadi Al Abyad the transition zone from mantle peridotites to gabbroic rocks of the lowermost oceanic crust (MTZ – Moho transition zone; Moho – Mohorovičić discontinuity) is exposed over a distance of several hundred meters (Fig. 146). At some places, the Moho is represented by a sharp planar contact between peridotite (harzburgite and dunite) and layered gabbro (Fig. 147). At a later stage this planar contact was faulted. The contact between peridotite and gabbro seems to be folded at some places (Fig. 148), which, however, may also be interpreted as a primary interfingering contact. The dunites contain plastically deformed thin sills of gabbro. Layering in the mantle material is in general parallel to the Moho. Near the boundary corresponding to the fossil Moho a number of layers with cumulate rocks occur. The cumulate rocks contain large crystals, which are partly more than 10 cm in diameter, of plagioclase and pyroxene (Fig. 149). Directly below the fossil Moho a cross cutting chromite vein can be observed at the outcrop shown in Fig. 147. The plutonic section of the oceanic crust starts with layered gabbros (Fig. 150) which make up the lower two-thirds of the gabbro layer of the oceanic crust. The layering is an effect of variations in the composition of

Fig. 148: Mixing of peridotite and gabbro at the fossil Moho in the Wadi al Abyad.

Fig. 149: Ultramafic cumulate with large crystals of pyroxene and plagioclase in a cumulate layer of the peridotite along the fossil Moho in the Wadi al Abyad.

Fig. 150: Layered gabbro in the Wadi al Abyad.

the gabbros regarding the proportion of plagioclase, clinopyroxene and olivine (MacLeod & Yaouancq 2000).

Although they are very rare, wehrlites as a special product of the lithospheric mantle can be observed in thin layers directly at the contact of peridotites and gabbros. Whereas the surrounding gabbros contain plagioclase, the wehrlite at this location is mainly composed of clinopyroxene without any plagioclase. Wehrlites are classically interpreted as components of the mantle which form as intrusives from the deeper lithopheric mantle into the tectonised, sub-Moho peridotites (Nicolas 1989). However, the wehrlites in the MTZ are different since they formed under low-pressure conditions. They mostly show a characteristic cumulate texture suggesting that these rocks were formed by the accumulation of olivine and clinopyroxene in shallow crustal magma chambers (Koepke et al. 2009).

EP 59

Tawiyah chromite
Topic: Podiform chromite in ultramafic rocks
Location: UTM 40 Q 578622 2587250 / N 23°23'35" E 57°46'09"

Rating: ☺

For location map see EP 58, Fig. 145.

Approach: Leave the main road #1 (A'Seeb Street) west of Muscat at Barka towards the south-southwest to Nakhl and follow road #13 for about 48 km, passing through Nakhl towards the west to Al Awabi. 4.5 km after Nakhl, turn right into a paved road to Tawiyah and leave this road after 500 m by turning left into an unpaved road which leads into an abandoned quarry. Part of the mined out quarry is refilled.

Small remnants of a chromite ore body are still observable in this abandoned and only partly refilled quarry (see EP 34 for a description of chromite ores). The best chromite outcrop is at an east-facing wall directly next to the water-filled part of the quarry (Fig. 151). The ultramafic rocks are completely serpentinised and large fiber crystals of chrysotile are abundant. The upper part of the ultramafic rocks is severely weathered. A regolith overlain by fluvial gravel deposits is visible in a profile at the wall left from the mining activity above the water filled part of the quarry.

Fig. 151: Remnants of a chromite ore body near Tawiyah.

EP 60

Al Awabi magnesite
Topic: Alteration of ultramafic rocks, magnesite
Location: UTM 40 Q 557862 2579350 / N 23°19'22" E 57°33'57"

Rating: ☺

For location map see EP 58, Fig. 145.

Approach: Follow the road #13 from Al Awabi to Nakhl. Leave the car on the left hand side of the road about 2.5 km after Al Awabi, facing in the direction of Nakhl, parking off the road in an unpaved space. Walk towards the north into a small wadi. Several white magnesite veins are easily accessible in the vicinity of the coordinates.

The road from Nakhl to Al Awabi follows a valley through the mantle section of the Samail Ophiolite. The rock sequence is severely weathered and forms the rugged dark-coloured terrain. Caused by the obduction of the ophiolite, the rocks are highly deformed and serpentinised and a large number of veinlets and fractures can be observed. Irregular shaped decimeter-scale encrustations of white magnesite are abundant and

Fig. 152: Microkarst features (rillenkarren) in magnesite veins in ultramafic rocks of the Samail Ophiolite.

clearly visible. The formation of magnesite is a product of the alteration of serpentinite (see EP 1 and EP 88 for a description of the alteration process). Outcropping surfaces of the magnesite show nice microkarst features such as rillenkarren (Fig. 152) or small hollows (kamenitza).

EP 61

Permian unconformity Wadi Bani Kharus
Topic: Angular unconformity, stromatolites
Location: UTM 40 Q 555820 2570063 / N 23°14'20" E 57°32'44"

Rating: ☺☺☺

For location map see EP 58, Fig. 145.

Approach: Leave the main road #1 (A'Seeb Street) west of Muscat at Barka and travel towards the south-southwest to Wasit, following the road #13 up to Al Awabi. In Awabi, turn left into the road towards the south continuing for about 11 km on the paved road in the Wadi Bani Kharus. If you have a 4-wheel vehicle, it is possible to drive directly into the outcrop. If you are travelling in a normal sedan car, park by the side of the road and walk for a few meters. The outcrop appears on the left (eastern) side of the gorge and can be climbed on a blocky talus up to the unconformity plane (10–15 m height). The best time to visit the outcrop is in the morning when the sun is still illuminating the wall.

At this outcrop the exposure of the angular unconformity between Late Precambrian and Permian sedimentary strata is of an extraordinary good quality (Fig. 153). Late Precambrian limestones are tilted by an angle of about 35° towards the northwest and are discordantly overlain by horizontally lying Permian sediments.

The sediments below the unconformity consist of partly cross-bedded Late Precambrian limestones and dolomites of the Kharus formation (Fig. 154). Most significant are frequently occurring stromatolites (Fig. 155) which indicate shallow and warm water conditions at that time. Stromatolites are bulbously growing bio-sedimentary structures which develop from bacterial mats. Recent stromatolites are mostly found in hyper-saline waters, in particular in lagoons, where grazing predators are excluded due to high salinity, thus enabling an unhampered growth. In the Proterozoic era stromatolites settled both in hypersaline shallow water and in waters without hypersaline conditions. The earliest stromatolites that are around 2.8 billion years old, were found in South Africa. At this outcrop an average age of 650 million years is assumed.

The angular unconformity represents an old surface which was eroded during the Late Palaeozoic when all continents formed the supercontinent Pangaea. Subsidence can be observed from the Permian onwards. This results in the sedimentation of a thick sequence of marine sediments and is a consequence of the opening of the Neotethys

Fig. 153: Angular unconformity between Late Precambrian and Early Permian sedimentary rocks at Wadi Bani Kharus.

Fig. 154: Detailed view of the angular unconformity between Late Precambrian and Early Permian sedimentary rocks at Wadi Ban Kharus.

Fig. 155: Late Precambrian stromatolites below the angular unconformity at Wadi Bani Kharus.

Fig. 156: Transgression conglomerate at the base of the Permian Saiq formation, unconformably overlying the Precambrian Kharus formation.

Ocean and the breakup of Pangaea. At that time the rocks forming the Jebel Akhdar Mountains today, were located at the south-western passive margin of the Neotethys Ocean separating the Cimmerian Superterrane from the African Plate (see Fig. 16). The Arabian Plate was then still connected to the African Plate.

The Permian sedimentary succession of the Saiq formation (262–250 Ma) begins with a transgression conglomerate at the base which is exposed at this location. The lowest conglomerate is composed of rocks erosionally derived from the Precambrian substratum; dolomites and some reworked stromatolites occur among red and black cherts and sandstone (Fig. 156). The conglomerate is overlain by sand- and siltstone and later by dolomite indicating the increasing subsidence at the passive margin.

EP 62

Wadi Haslan, Snowball Earth hypothesis
Topic: Neoproterozoic glaciations, tillites and cap carbonates, Snowball Earth hypothesis
Location: UTM 40 Q 550608 2565506 / N 23°11'52" E 57°29'40"

Rating: ☺☺☺

For location map see EP 58, Fig. 145.

Approach: Leave the main road #1 (A'Seeb Street) west of Muscat at Barka towards the south-southwest to Wasit and follow road #13 to Awabi. In Al Awabi, turn left onto the road towards the south continuing for about 8 km, then turn right (at UTM 40 Q 553860 2571155 / N 23°14'56 E 57°31'35"). After 3 km the outcrop appears on the western side of the Wadi Haslan at a hill of approximately 10 m height about 150 m away from the newly constructed main road which was completed in 2012. There is a footpath leading directly into the outcrop from the south. Please respect the fences of the owner of the farm on top of the hill.

The hill is composed of a diamictite which is unconformably overlain by a dolomite close to the top. According to the geological map of Beurrier et al. (1986) the sequence is of Neoproterozoic age and belongs to the slightly metamorphosed Mistal formation (tillite: Mi1; dolomite: Mi2D). In more recent publications the diamictite is assigned to the Huqf supergroup with the diamictite as Fiq formation and the cap dolomite as Hadash formation (Leather et al. 2002, Kilner et al. 2005). The age of the Huqf supergroup is determined by U-Pb zircon age dating at 711.8 million years from an ash unit in the lower part of the group (Allen et al. 2002). This relates the rocks of this locality to the Sturtian glaciation during the Cryogenium of the Neoproterozoic (see Fig. 22).

The dark brown weathered lower part is made up of a matrix supported grey diamictite which most probably formed as a tillite. It contains numerous angular components up to 15 cm in size. Some of them show striations on their surface which are an indica-

Fig. 157: Neoproterozoic dolomites ("cap carbonate sequence") on top of Neoproterozoic tillites at Wadi Haslan.

tion of ice transportation. The overlying light brown dolomite with layers of about 2–3 cm thick is supposed to have formed in a shallow marine depositional environment indicating warm climatic conditions (Fig. 157). The question how the conditions may change within a very short time from icehouse to greenhouse without a large hiatus between has triggered a lively discussion (see Hoffman et al. 1998, and references therein). The area of Oman which geologically belongs to the Arabian Platform has been proposed to be situated at low palaeo-latitudes in the Late Neoproterozoic (Kempf et al. 2000, Kilner et al. 2005). This situation – both, glacial sediments and unconformably overlying carbonates without a large hiatus – supports the so-called Snowball Earth hypothesis (cf. chapter 5.3.2, Excursus II: Snowball Earth – the largest glaciation of the Earth's history).

The term "Snowball Earth" was introduced by Kirschvink (1992) who claimed that planet Earth was completely covered by ice sheets in the Neoproterozoic. A reason for this is seen in the break-up of the supercontinent Rodinia. Here, large areas were desert-like and dry over a long time, because of their large distance to the open sea. These regions were again affected by the environmental effects of rainfall and thus chemical weathering due to the break-up. This led to the elimination of carbon dioxide from the atmosphere and consequently to decreasing temperatures. The global glaciation caused a freezing of the oceans probably down to a depth of 2 km. Additionally, the self-energising albedo effect contributed to the global glaciation by reflecting the incoming sunlight with large areas covered by ice, allowing less thermal energy to reach the Earth´s surface. However, outgassing of CO_2 caused by continuous volcanic activity finally led to a strong increase in CO_2 in the atmosphere and a change to warm-house conditions within a short time span.

Similar rock sequences, capped by dolomite with tillites at the base exist at several locations around the world.

EP 63

Pass to Wadi Bani Awf
Note: a 4-wheel vehicle is recommended for this excursion.
Topic: Neoproterozoic sediments, stromatolites, good wadi views
Location: UTM 40 Q 548152 2566860 / N 23°12'36" E 57°28'14"

Rating: ☺

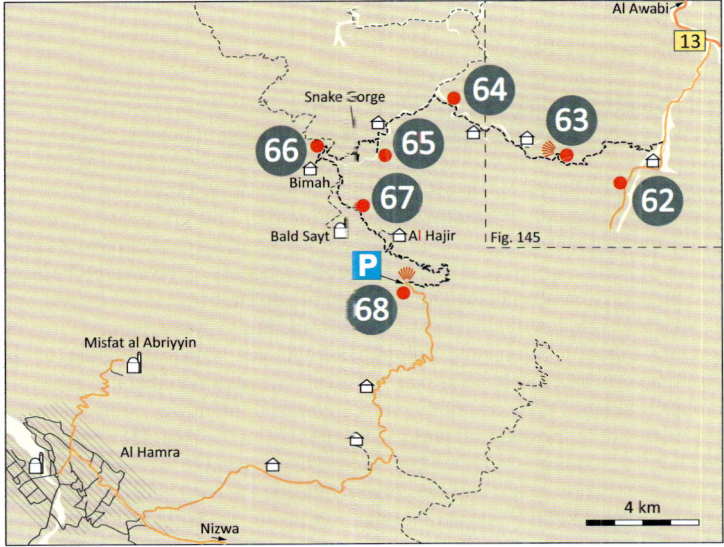

Fig. 158: Location map.

Approach: Leave the main road through Wadi Haslan (see EP 62) at 40 Q 552116 2567406 / N 23°12'53" E 57°30'33" and turn uphill onto the unpaved road. You will reach the pass after about 6 km with a nice viewpoint of the view to the west into Wadi Bani Awf. The outcrop appears directly along the road after a sharp curve towards right.

The stop is located in the center of the mountains, which can be described as a basement cored anticline. Exposed here are late Proterozoic layered limestone and dolomites. Stromatolites can be observed. These are layered accretionary structures. Microbial mats are important in the formation as they trap fine grained suspended sediments. By this process the stromatolites grow, commonly in a tidal-flat environment. The rocks are slightly metamorphosed and tilted with open folds. The stop further allows a nice view into Wadi Bani Awf and to the adjacent mountain tops.

EP 64

Wadi Bani Awf – pencil cleavage
Note: A 4-wheel vehicle is highly recommended for this excursion.
Topic: Neoproterozoic rocks, Muaydin formation, pencil cleavage
Location: UTM 40 Q 545169; 2572305 / N 23°15'33"; E 57°26'29"

Rating: ☺

For location map see EP 63, Fig. 158.

Approach: Numerous outcrops occur along the road from Wadi Haslan (see EP 60) into Wadi Bani Awf. A well exposed example is at the location indicated.

The central parts of the Al Hajar Mountains are made up of Precambrian rocks, as the entire structure represents a basement-cored anticline. The pre-nappe autochthonous unit (Muaydin formation) exposed along the track here is characterised by mauve to green siltstone with thin beds of sandstone and carbonate. They were deposited in a littoral sedimentary environment. The rocks are nicely stratified and the bedding planes intersect at a high angle with the cleavage. Cleavage is a tectonic planar fabric and type of secondary foliation. Elongated polygonal fragments with rectangular to rhombic cross sections form as a result. They can be observed as weathered material all along the slopes. The term *pencil cleavage* or *Griffelschiefe*r in German was coined in areas dominated by slate where the material was used as pencils in the past. However, the pencils observed here are commonly rather large and rather resemble swords or daggers. The Arabic term for sword is *seif* and the local name for the rock fabric is therefore *seif-cleavage*, or *Schwertschieferung* in German. Large sigmoidal quartz veins occupying tension gashes can also be observed in the surroundings.

EP 65

Snake Gorge, near Bimah
Note: A 4-wheel vehicle is highly recommended. It is also recommended not to hike in the gorge after rain or if there is a possibility of rain.
Topic: Erosion, narrow gorge, Permian limestone
Location: UTM 40 Q 541293 2567139 / N 23°12'46" E 57°24'12"

Rating: ☺☺

For location map see EP 63, Fig. 158.

Approach: Continue on the unpaved road from Wadi Haslan (see EP 63) into Wadi Bani Awf and turn to the left at the junction at 40 Q 543505 2569166 / N 23°13'51" E 57°25'23". Turn left after about 3.3 km and go through some small villages. Continue to drive in the wadi for about 400 m. The outcrop is at the end of the road.

The Snake Gorge is an impressively deep valley which is a branch of the Wadi Bani Awf. It is apparently named for its snake-like winding and narrow nature, forming a ravine of 5 kilometers length. Due to its narrow nature, the water flow after rain and storms are usually very intense. Because there is hardly any vegetation or soil to trap the water, it runs off the bare rock very quickly into the wadis causing a flash-flood.

Fig. 159: Snake Gorge cut into dark limestones of the Hajar formation.

Geologically, the gorge is located at the southern rim of the Sahtan bowl, an anticlinal structure in the Al Hajar Mountains which is part of the Jebel Akhdar mountain range. Here, Late Proterozoic rocks are exposed in the inner part of the dome. A folded sequence of the autochthonous series of the Oman Mountains is exposed along the way through the Al Hajar Mountains, which is the only way to pass this high mountain range. At the Snake Gorge, the rocks belong to the Hajar formation in the upper part and to the Muaydin formation in the lower part. The Hajar formation consists of fetid black dolomitic limestone with some algal limestones and small stromatolitic structures which are interpreted as formed in a lagoonal environment. The very narrow gorge at this location (Fig. 159) is cut into the dark limestones of the Hajar formation.

EP 66

Snake Gorge view from top
Topic: Viewpoint, erosion
Location: UTM 40 Q 540363 2566980 / N 23°12'41" E 57°23'40"

Rating: ☺☺

For location map see EP 63, Fig. 158.

Approach: Continue on the unpaved road from the Snake Gorge exit (see EP 65) to Wadi Bani Awf and turn to the left at the junction at 40 Q 541307, 2567541 / 23°12'59" N, 57°24'13" E. Drive uphill on the dusty track and park the car on the side of the track somewhere around the given co-ordinate points.

This location offers a view into the narrow valley of the Snake Gorge from above. It becomes evident that the incision that led to the formation of the small but steep canyon must have been fast. Otherwise the valley formation would have been V-shaped. The reasons for this fast incision can have (a) climatic reasons as, e.g., more precipitation leads to more erosion; can be (b) controlled by oscillations of the base level. However, eustatic sea-level changes seem unlikely here as too far from the sea. Furthermore, the reasons can be (c) the result of recent crustal uplift. But, the most likely explanation is (d): a combination of a–c.

The canyon is popular with adventure-seeking hikers as the gorge is so narrow that parts of it have to be tackled by a combination of swimming and jumping. Tragically, nine people drowned in the wadi following a sudden downpour and precipitation event in 1996. The hikers were surprised by the "speed how the situation could so quickly and so dramatically change." You can find the testimonial of Chris Marsden who survived the event here: http://home.kpn.nl/lilian_jan_schreurs/oman/Marsden.htm.

EP 67

Thrust fault Al Hajar
Note: A 4-wheel vehicle is highly recommended.
Topic: Ramp structure, thrust fault
Location: UTM 40 Q 540636 2564901 / N 23°11'33" E 57°23'49"

Rating: ☺☺

For location map see EP 63, Fig. 158.

Approach: Continue on the unpaved road from Bimah (see EP 66) to Nizwa crossing the Al Hajar Mountains range. You will reach the viewpoint about 3.2 km after EP 66. The view of the outcrop from a distance is on the left hand side, northeast of the road on the other side of the deep wadi.

The Muaydin formation belongs to the Late Proterozoic rocks which are exposed in the Sahtan bowl. It occurs below the Hajar formation and is mainly composed of slightly metamorphosed green siltstone and thin beds of sandstone, reflecting a littoral environment. These rocks are multiply deformed with two more or less rectangular cleavage planes resulting in well-developed pencil structures in the schists.

A good clear view of a folded and thrusted ramp structure within the dolomite of the Hajar formation can be seen on the road passing the Al Hajar Mountains above the Wadi al Hat. It is an excellent example of thrusting (Fig. 160).

Fig. 160: Thrusted ramp structure in dolomites of the Hajar formation.

EP 68

Viewpoint into Sahtan bowl
Topic: Geomorphological overview
Location: UTM 40 Q 542604 2562132 / N 23°10'03" E 57°24'58"

Rating: ☺☺

For location map see EP 63, Fig. 158.

Approach: The site is located on a mountain pass and therefore two alternative routes are possible. The access from the north is via the well worthwhile drive through Wadi Bani Awf (see EP 63 through EP 67) This drive is only possible with a 4-wheel vehicle. The southern access road is paved and follows the dip of the Cretaceous limestone of the Kahwah group. Take the Ar Rawdah – Al Hamra Road at the foot of the mountain and follow the road up at the intersection at UTM 40 Q 532603 2553164 / N 23°05'12" E 57°19'06". Follow the road for 23 km to reach the viewpoint.

Northern Oman is dominated by the mountain chain known as either Oman Mountains or as Al Hajar Mountains. Jebel Al Hajar translates to 'stone mountains'. This viewpoint is located around 2,000 m above sea level and allows a fantastic view into the mountains, or – if the visibility is good – even up to the Batinah coastal plain and the coast. The parts of the mountains visible form here include Wadi Sahtan to the northwest, Wadi Bani Awf to the north-northwest, Wadi Bani Kharus to the north-northeast. The Gubrah Bowl can be seen in the distance to the northeast.

Extensive weathering and subsequent erosion has partly stripped off the Mesozoic sequence to the extent that the Precambrian basement is exposed (see EP 54, 55 and EP 61 through EP 64). The overlying nappe deposits are eroded as well and the structures therefore represent geological windows. The sediments have been transported into the foreland to form bajadas (merged alluvial fans) both in the north i.e. the Batinah coastal plain, as well as south of the mountains. The dark coloured rocks directly north of the viewpoint are Precambrian, whereas the viewpoint itself is located on top of Cretaceous clayey limestone which is exposed in the road section next to the parking lot.

EP 69

Hasat Bin Salt, Al Hamra
Topic: Archeological site, Coleman´s rock, carvings, Neolithic
Location: UTM 40 Q 528975 2551802 / N 23°04'28" E 57°16'58"

Rating: ☺☺

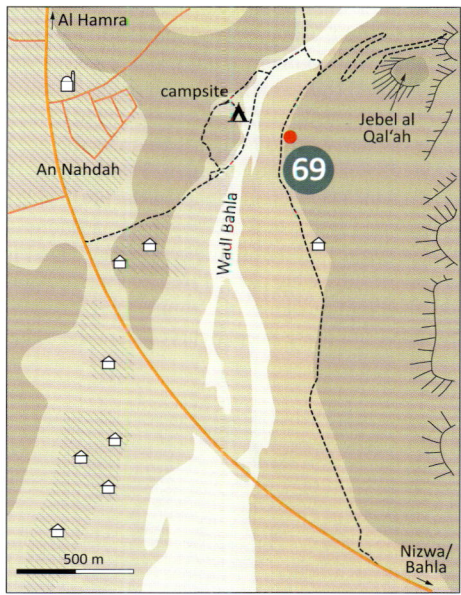

Fig. 161: Location map.

Approach: Take the main road from Al Hamra to Nizwa. Leave the road behind the village of An Nahdah at UTM 40 Q 528083 2551247 / N 23°04'10" E 57°16'26" and continue on an unpaved road towards the north-east running parallel to the Wadi Bahla. In the background you can easily recognise a steep cylindric mountain which forms a characteristic landmark. This landmark named Jebel al Qal'ah is near the archeological site.

Hasat Bin Salt or Hasat Bani Salt, also known as Coleman's Rock, is a limestone-boulder of 6 meters high described in detail by Yule (2001) and Reade (2000). It is located on the wadi bed in the shade of the impressive landmark of Jebel al Qal'ah. Hasat

Bin Salt means "rock of Bin Salt" – named after a tribe which lives in Wadi Ghul, located north-east of this place. The boulder took its second name "Coleman's Rock" from the geologist Robert Coleman who made the rock well-known in the 1970s. The boulder represents the most important rock art monument of southeastern Arabia (Fig. 162). The different groups of figures can be best seen at different times of the day, depending on the illumination. However, the carvings are best studied using artificial light during the night.

The south-southwest and north-northwest face of the rock are carved in low relief showing seven human figures: males, females and one child. The figures are visible best in morning- and sunset light. The group on the south wall becomes slightly better observable in the morning. This southern main group consists of a large man (2.5 m high) in the centre with a raised right arm holding an object in his hand. A woman, easy to recognise by her indicated breasts, stands to his right. There is another, smaller individual, probably of male gender, next to her. This figure is the one of poorest preservation in this group. On the other side of the central male individual, there is the only child showing an emphasised navel. The second group on the north side consists out of three figures, probably one man and two women. They are in a poorer state of preservation than the main group. It is suggested that the reliefs were carved into the walls in two phases. The women on both walls were carved first and the remaining figures followed later. Indicators are the different design of the figures and the different carving style – the younger four individuals are more deeply carved than the older ones. The tall male

Fig. 162: Coleman's rock: Four life-size figures are visible on the south face.

figure on the south wall is even crowding the woman; he was probably carved over the original companion of the woman.

Further investigations of the rock in 2012 by one of the authors (GH) have revealed the existence of additional two figures on the western side of the boulder. Here one figure – probably male – in full body size as well as one smaller head can be identified around the middle of the day when the sun illuminates this rock face.

Age and purpose of the reliefs are unknown. Almost certainly, they are pre-Islamic as the depiction of female's breasts is improper according to Islamic principles. They might be as old as 4,000 years (Cleziou & Tosi 2007d). Although rock art is common in Oman's wadis, the style and size of the figures on Hasat Bin Salt are nearly unique.

EP 70

Falaj Daris Nizwa
Topic: Irrigation systems
Location: UTM 40 Q 556319 2540625 / N 22°58'22" E 57°32'57"

Rating: ☺☺

Approach: Follow the main road through Nizwa towards the northern end of the town. Before leaving the town, turn right at a newly constructed roundabout. The location is signposted. You will enter a recreation park (Merfa Dares Park) with a carpark.

Under the name "Aflaj Irrigation System of Oman", the Falaj Daris, together with four other locations, became listed as UNESCO World Heritage sites in 2006 (UNESCO 2014b). The term falaj (plural aflaj) is classical Arabic and means "to divide property". Here, the term is applied to an irrigation system which divides water among stakeholders. Physically, it is a system that collects, channels and distributes water for agriculture or domestic purposes by using gravity. Directed by astronomical observations and moon phases, the water is fairly and effectively shared among the stakeholders. The importance of the systems as a main artery for villages and towns is indicted by numerous watch towers in the vicinities. They were built to defend the resource which was so essential for the survival of the people.

Three different falaj-types are documented:
- Daoudi Aflaj, which collects groundwater,
- Aini Aflaj, which collects spring water, and
- Ghaili Aflaj, which collects surface water.

Ghaili is the most prevalent followed by Aini and Daoudi. The principle of the aflaj dates back to at least CE 500. They might have existed even as early as 2,500 BCE but the historical evidence is uncertain as no written documentation exists. The aflaj system is seen as one of the driving forces behind the formation of Oman as a nation as it provided the reason for nomadic societies to settle down.

Field sites

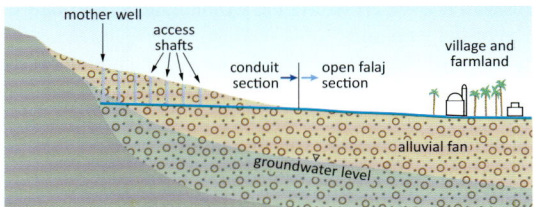

Fig. 163: Principle sketch of a Daoudi Falaj (simplified after Ministry of Regional Municipalities, Environment and Water Resources 2006).

About 3,000 such systems are still in use in Oman. The Falaj Daris in Nizwa is probably one of the oldest falaj of the Daoudi type in the Sultanate. The Daoudi Falaj was named after the Prophet Suleiman bin Daoud (Solomon, son of David) who, according to the legend, told a djinni to dig the falaj. In order to build a Daoudi Falaj without supernatural power, firstly the location of the mother well (see Fig. 163) needs to be determined. This step needs considerable experience. The upper slope of the alluvial fan as well as changes in vegetation and surface discharge are observed and considered. The well is sunk into the aquifer then and might reach depths of up to 60 m. It is left for a couple of days in order to prove a consistent water supply. The well-sides are stabilised with rocks or burnt bricks and the underground conduit is dug. About 0.9 m wide and 1.5 m high, this delivery channel might stretch over several kilometers. In unconsolidated sediments, support is provided by different materials like rocks, bricks and palm tree trunks. Access shafts are dug in regular intervals of around 20 to 100 meters each. They allow maintenance work like cleaning and reparation. The Falaj Daris in Nizwa consists of three channels which add up to a total length of 7,990 m (Ministry of Regional Municipalities, Environment and Water Resources 2006). The falaj is mainly fed by water from Wadi al Abyad. The flow rate reaches 2,000 l/s (Ministry of Regional Municipalities, Environment and Water Resources 2006) but the falaj is affected by over-exploitation and the flow rate drops significantly during times of droughts.

EP 71

Corals on the Saiq Plateau
Note: Only 4-wheel vehicles are allowed to pass the police checkpoint at the foot of Jebel Akhdar.
Topic: Coral Garden and Al Noor viewpoint in Al Jebel al Akhdar
Location: UTM 40 Q 568566 2551155 / N 23°04'03" E 57°40'09"

Rating: ☺

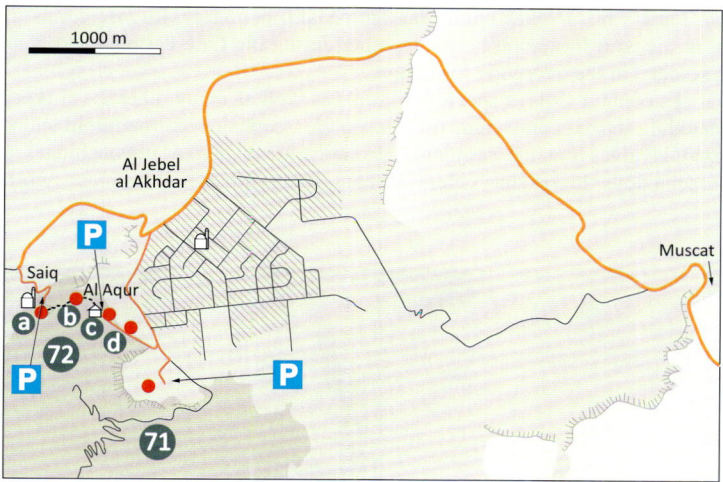

Fig. 164: Location map.

Approach: Drive to Birkat al Mouz on the road from Muscat to Nizwa. Follow the way to the village of Al Jebel al Akhdar through Wadi Muaydin (see also Fig. 168). Starting at the police station (ROP) at UTM 40 Q 569516 2539341 / N 22°57'39" E 57°40'40", follow the road for about 32 km to reach the Saiq Plateau and continue through Al Jebel al Akhdar into the little village of Al Aqur (UTM 40 Q 568237 2551714 / N 23°04'21" E 57°39'58") which is located directly below Al Jebel al Akhdar at the end of the paved road going down at the rim of the plateau.

The coral garden of the Saiq Plateau is situated in a very attractive area frequently visited by tourists to enjoy the scenic viewpoints of the garden terraces and mountain villages in Al Jebel al Akhdar. At this location a large variety of well-preserved marine fossils can be studied at an elevation of more than 2,000 m above sea level. The rocks contain fossils that date back to Permian times, about 250 million years ago.

Solitary and colonial corals abundantly occur in the lowest part of the Saiq formation (Khuff) which was deposited in a wide carbonate platform that covered most of the Arabian Plate during Late Permian. This carbonate platform developed at a passive continental margin caused by the spreading of the Neotethyan Ocean and as a consequence of the breakup of the supercontinent Pangaea. The Saiq formation rests with an angular unconformity on the pre-Permian sedimentary basement (carbonate and glacial rocks from the Precambrian, cf. EP 61) which mirrors a large-scale regional transgression phase that occurred during the Permian. It is conformably overlain by the Mahil formation. The Saiq formation was deposited during a period of relative tectonic quiescence. The boundary between Permian and Triassic, which marks the largest extinction phase in Earth's history lies within the Saiq-Mahil boundary. Rugose corals that are present in the Saiq formation became extinct during the Late Permian and, therefore, cannot be cited in the overlying formation.

The rocks contain various species of corals as well as many other fossil types (Fig. 165) including inozoan and sphinctozoan sponges, cephalopods (nautiloid and ammonoid taxa), bivalves (e.g. the Permian extinct gigantic bivalve Alatoconchidae), gastropods, brachiopods, bryozoans, echinodermata, foraminifera and different types of al-

Fig. 165: Late Permian, Saiq Plateau, Jebel Akhdar; a: plan view of fasiculate coral; b: fasiculate coral head in life position; c: ammonite; d: detail of weathered fasiculate coral.

Fig. 166: Trace fossils in Permian limestore of the Saiq Plateau.

gae. The coral garden includes corals that can be differentiated to solitary (horn-shaped single coralites), fasciculate (colonial open-branching) and thamnasterioid (colonial massive compound).

The solitary corals might have formed in a deeper and cooler water environment below the photic zone. The open-branching type of corals might have formed in high energy zones, whereas the massive compound ones probably grew in low water energy settings. Rugose corals are the most dominant and most important coral type in the coral gardens. These corals lived from Middle Ordovician and became extinct in Late Permian. They inhabited shallow marine warm water reef environments. Other types of corals include tabulate corals that lived from lower Ordovician to Permian. These exclusively occur in colonial forms that have slender coralite tubes which are commonly closely packed and show circular, oval or polygonal shapes in transverse sections.

Some layers show intensive bioturbation (Fig. 166). These appear as spreiten from *Zoophycos* (Knaust 2009).

EP 72

Tufa on the Saiq Plateau
Topic: Karstification, tufa formation, terraces, walk through villages
Note: Only 4-wheel vehicles are allowed to pass the police checkpoint at the foot of the Jebel Akhdar.
Location: UTM 40 Q 567957 2551822 / N 23°04'25" E 57°39'48"

Rating: ☺

For location map see EP 71, Fig. 164.

Approach: Drive to Birkat al Mouz on the road from Muscat to Nizwa. Follow the road to the village of Al Jebel al Akhdar through Wadi Muaydin (see also Fig. 168). Starting at the police station (ROP) at UTM 40 Q 569516 2539341 / N 22°57'39" E 57°40'40",

Fig. 167: Waterfall over the edge of the Saiq Plateau following a thunderstorm. The terraced gardens are located on a giant tufa deposit.

follow the road up for about 32 km to reach the Saiq Plateau and continue through Al Jebel al Akhdar on the road to the airport. About 1.5 km after the village of Al Jebel al Akhdar, turn left into a small road to the village of Saiq. After about 500 m you will reach a carpark (UTM 40 Q 567828 2551818 / N 23°04'25" E 57°39'44"). Continue from here on foot.

The stop is located on the edge of the Saiq Plateau at an elevation of around 2,000 m. Here, the annual precipitation is the highest measured in northern Oman and amounts to more than 300 mm/year (Kwarteng et al. 2009). Following precipitation events, water gushes down the slope to form an impressive waterfall (Fig. 167).

There are three villages seemingly glued to the edge of the plateau: Sharijah, Al Ayn and Al Aqur. The mountain front facing Wadi Muaydin is almost vertical. Nevertheless, over the centuries the villagers have transformed this slope into hanging gardens by terrace farming. A sophisticated system of irrigation channels (falaj) enables the agricultural use. The terraces follow the mountain face like contour lines on a map. You can hike along a sign-posted trekking path through the villages, gardens and orchards. The gardens are an example of a unique high altitude agro-ecosystem that includes the cultivation of pomegranate, walnut, apricot, almond and peach (Buerkert & Schlecht 2010). The area is also famous for the production of Omani rosewater and rose oil, distilled from the damask rose *Rosa* x *damascene* that grows on the terraces as well.

The plateau itself is made up of Late Permian limestone and dolomite (cf. EP 71). These rocks are subject to intense weathering and karstification. The latter being the result of the dissolution of soluble rocks by acidic water. Common features of karst landscapes include caves and sinkholes (cf. EP 10 and EP 12). Here at this stop, rapid precipitation of calcium carbonate in the form of tufa can be observed. Tufa is defined as a soft, terrestrial, porous freshwater $CaCO_3$ deposit. See Ford and Pedley (1996) for a general discussion on the difference between tufa and travertine. The whitish tufa deposits form spectacular carbonate precipitates that encrust the pre-Permian formations over a height of 400 m. The tufa cone resembles a giant petrified waterfall when seen from the distance (e.g. from EP 71). The process of tufa formation here is the result of water turbulence as the gradient steepens at the edge of the plateau into Wadi Muaydin. The accelerated flow velocity leads to an enlargement of the air-water interface, allowing an increase in CO_2 outgassing which in turn leads to an increase in the pH value. This leads to calcite super-saturation. Consequently, calcite precipitation is induced as the carbonate solubility decreases (Zhang et al. 2001).

EP 73

Mesozoic carbonate platform in Wadi Muaydin
Topics: Mesozoic carbonate platform, petroleum fields
Location: UTM 40 Q 568357 2539748 / N 22°57'52" E 57°40'00"

Rating: ☺☺

Approach: Take the road #15 from Muscat to Nizwa. Just before arriving in Nizwa, on the main dual carriageway, turn right to Birkat al Mouz (UTM 40 Q 570226 2531589 / N 22°53'26" E 57°41'01") and follow the signs to Al Jebel al Akhdar (Saiq Plateau) and Wadi Muaydin. Before ascending on the mountain road, take a left turn on a graded road (UTM 40 Q 568848, 2538088 / N 22°56'58" E 57°40'17"), heading inside Wadi Muaydin (see Fig. 168). Drive for 2.2 km. The outcrop is on the left side.

In this outcrop the contact between the Early and Late Cretaceous carbonate platforms is exposed. Driving inside the wadi, the Jurassic rock units follow at UTM 40 Q 568801 2541901 / N 22°59'02" E 57°40'17" and the Triassic carbonate layers at UTM 40 Q 568846 2542467 / N 22°59'20" E 57°40'18", where the village of Al Muaydin is located at the end of the road. The wadi also forms an incision through the allochthonous, overthrusted rocks of the Hamrat Duru group that were deposited in the Neo-tethys Ocean (cf. EP 74). They are highly deformed with massive folds and faults that were mainly formed as the rocks were being transformed from the ocean to land.

While driving to Wadi Muaydin, you pass the famous Fort Bait al Rudaidah (cf. EP 74) and the UNESCO-registered water channel (falaj) of the Al Khatmayn next to it.

During Mesozoic times, massive carbonate platforms covered the eastern part of the Arabian Plate. The rocks deposited on these platforms cover great parts of the Al Hajar Mountains. One of the best inlets or wadis to examine is Wadi Muaydin, which is located on the southern flank of the Jebel Akhdar Mountains in the Wilayat of Nizwa. It discharges to the south where it forms a huge alluvial fan.

The platform edge of the Mesozoic carbonates was located just north of the present-day Oman Mountains, in the north of Oman (Droste & Steenwinkel 2004). Pulses of clastic sediment onto this platform were derived from the exposed Arabian Shield and fringing exposures of Palaeozoic sediments in the southwest. During Cretaceous, this platform was initiated in central Oman following a major transgression over the base Cretaceous unconformity (Fig. 169). After a rapid progradation of around 250 km to the north and north-east from Rayda to Kharaib, the platform edge aggraded (from Shuaiba to Natih times), leading to the development of a 700 m thick and 1000 km wide platform succession (Fig. 170).

The Cretaceous carbonate platform of Oman shows a complex internal architecture, rather than a "layer-cake" configuration. For example, the Habshan formation of the Early Cretaceous shows large-scale Arabian Plate margin configuration, large clinoforms of approximately 300 m thick, whereas the Shuaiba and Natih formations show

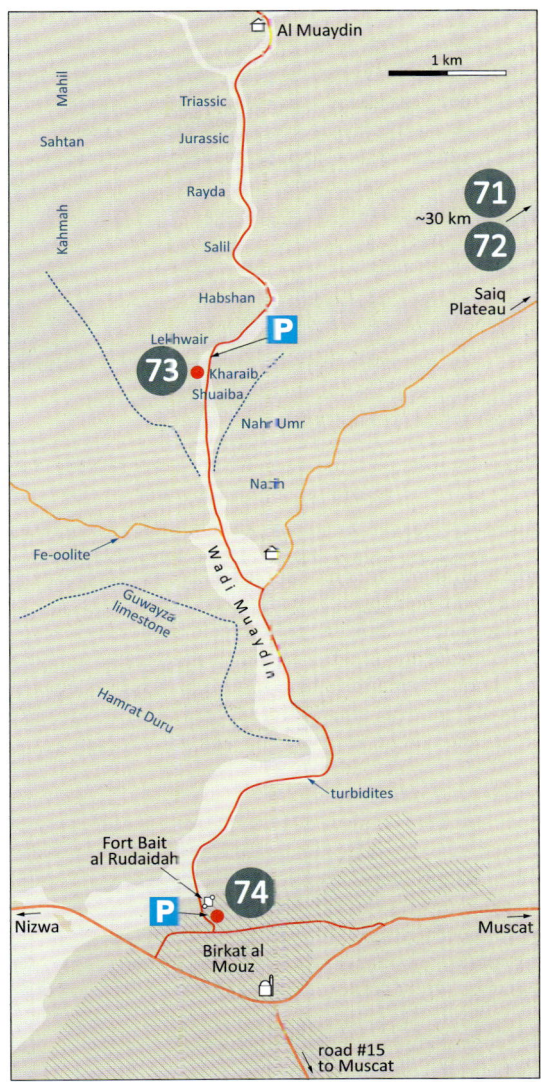

Fig. 168: Location map. Position of stratigraphic units (cf. Fig. 169) after Hanna (1995).

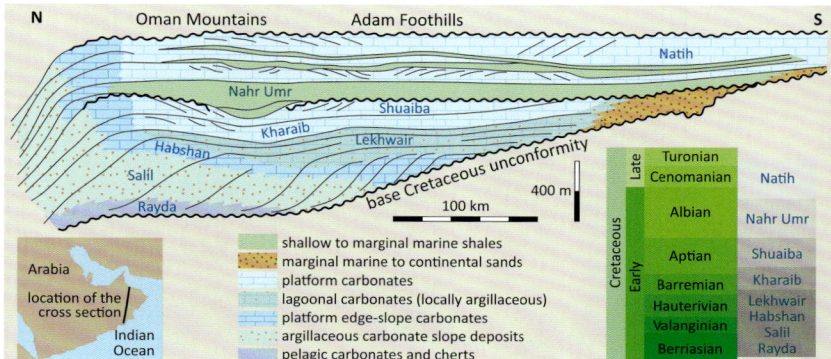

Fig. 169: Schematic sketch of the Cretaceous carbonate platform of northeastern Oman (simplified after Droste & Steenwinkel 2004).

different-scale intra-shelf basins. Smaller-scale clinoform complexes are associated with the margins of these intra-shelf basins. These intra-shelf basins are often filled with argillaceous lime mudstones interbedded with deeper-water shales and redeposited grainstones and packstones. The Mesozoic carbonate rocks form the main reservoirs in northern Oman and Arabia.

There are billions of barrels-worth of oil within these rocks, hundreds of meters below the ground in petroleum fields like Fahud, Natih, Lekhwair and Yibal in northern Oman. The Jurassic rocks in Wadi Muaydin consist of clastic sequences in the lower part, passing into cleaner carbonate to the top. Their interpreted depositional environ-

Fig. 170: Panoramic view of the Cretaceous carbonate platform sediments of Natih, Shuaiba, Kharaib, and Lekhwair formations in the Wadi Muaydin. View is toward WSW at EP 73 (see Fig. 168).

ment ranges from coastal estuarine and tidal flat to marine lagoonal and shoal environments (Bendias et al. 2015). The deposits of the Triassic Mahil/Sudair formation represent shallow water dolomitic carbonates with restricted faunas.

EP 74

Birkat al Mouz
Topics: Hawasina nappe, folds in turbiditic rocks; interesting, deserted old town
Location: UTM 40 Q 568394 2535240 / N 22°55'25" E 57°40'01"

Rating: ☺

For location map see EP 73, Fig. 168.

Approach: Take the road #15 from Muscat to Nizwa. Just before arriving in Nizwa, on the main dual carriageway, turn right to Birkat al Mouz (UTM 40 Q 570226 2531589 / N 22°53'26" E 57°41'01") and follow the signs to Al Jebel al Akhdar (Saiq Plateau). You will find the small outcrop at a wall next to the Al Yaariba Mosque at the northern end of the village just beside the Fort Bait al Rudaidah.

The name *Birkat al Mouz* translates to 'pool of bananas' and already indicates that the town is one of Oman's mountain oases. The modern houses are located along the main roads, whereas the historical deserted old town with adobe houses is located within lush green date palm plantations. The old town is well worth a visit. The gardens are fed by the 2.5 km long Falaj Al Khatmayn. It is a Daoudi Aflaj (cf. EP 70) that taps the water from the adjacent Wadi Muaydin with the mother well at UTM 40 Q 569288 2536777 / N 22°56'15" E 57°40'34". The falaj surfaces just north of the 17th century Fort Bait al Rudaidah, located at the entrance to the wadi.

The small outcrop of interest is located by the parking lot at the base of the Al Yaariba Mosque. Exposed here are lithoclastic limestone and chert layers that are part of the allochthonous Hawasina nappe and are mapped as Cenomanian (Hutin et al. 1986). Intense disharmonic folding with congruent and parallel folds, fold collapse in the hinge zone and thrusting can be observed in this outcrop. The deformation is caused by the emplacement of the nappe.

EP 75

Wadi Ghul
Topics: Cretaceous limestone, rudists
Location: UTM 40 Q 520875 2560070 / N 23°08'57" E 57°12'14"

Rating: ☺

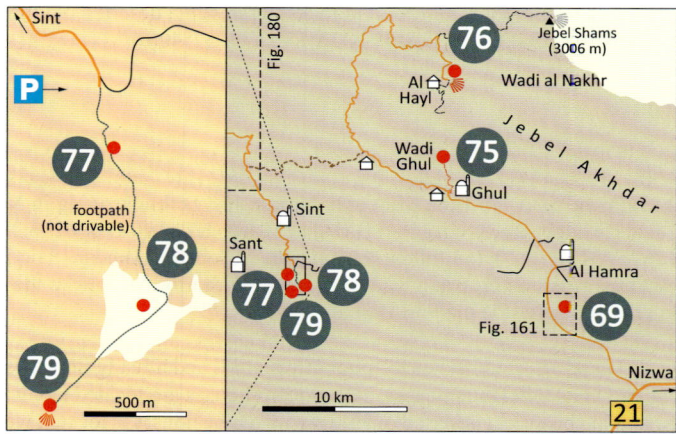

Fig. 171: Location map.

Approach: Leave the main road #21 from Nizwa to Bahla at the northern entrance of Bahla towards the north-west in the direction Al Hamra. Turn left at the petrol station in Al Hamra and follow the road for about 9 kilometer towards the north-west. The stop is a viewpoint overlooking the abandoned historic city of Ghul. You can turn right into an unpaved road leading to the Wadi Ghul here.

Wadi Ghul, also referred to as Wadi al Nakhr, is one of the deepest wadis of the Jebel Akhdar Mountains. The wadi drains along the southern flank of the giant anticline structure that forms the mountains. The strata are therefore inclined towards the south here. The modern houses of the village of Ghul are located on the right hand side of the wadi, whereas ruins of houses and fortifications are visible on the left hand side. These structures were built from the local rocks which make them camouflaged on the slope. The dip-slope they are standing on is composed of clayey limestone of the Cretaceous Fitri formation which disconformably overlies the Natih formation.

Upon entering the wadi, you will pass through a series of Neotethyan limestones that formed as platform carbonates on the passive margin of the Arabian Plate. These are

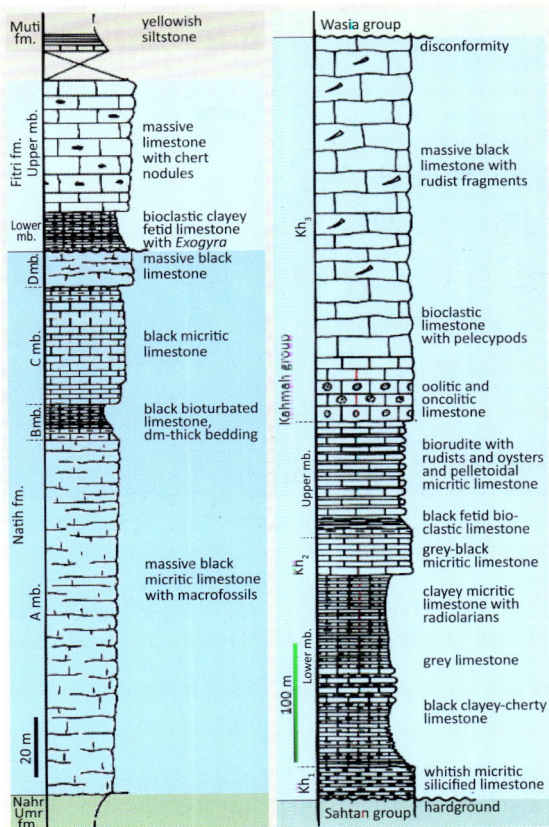

Fig. 172: Left: columnar section of the formations at the entry to Wadi al Nakhr. Right: columnar section of the Kahmah Group in the Wadi al Nakhr (modified after Beurrier et al. 1996).

Cretaceous in age at the entry to the wadi and Early Jurassic (Sahtan group) towards the end. The section starts with poorly stratified metre-thick beds of massive limestone of the Natih formation (see Homewood et al. 2008, for a detailed description) that crop out with no apparent sedimentary break on top of the the Nahr Umr formation at UTM 40 Q 520947 2560792 / N 23°09'21" E 57°12'16" (Fig. 172 left). First Jurassic rocks are exposed along the wadi bed at approximately UTM 40 Q 520677 2561549 / N 23°09'45" E 57°12'07". These highly bioclastic black oolitic limestone rocks (Fig. 172 right) con-

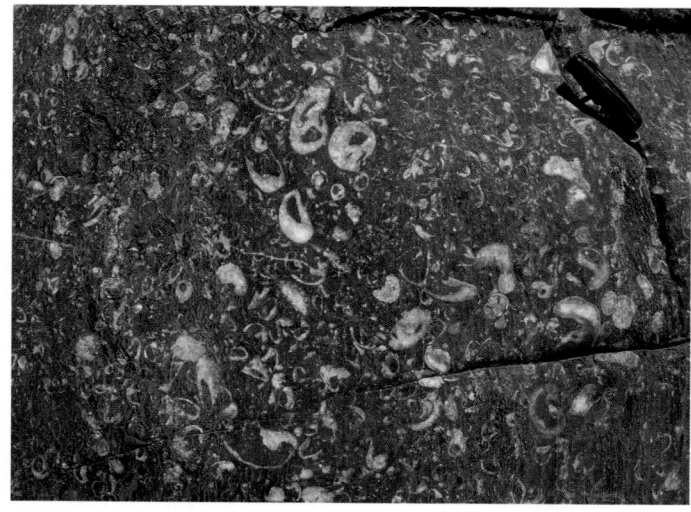

Fig. 173: Dark limestone with numerous rudist fossils of Hippuritoidae, Jurassic.

tain rudist fragments (Fig. 173). A conjugate fault system can be observed at UTM 40 Q 520892 2563592 / N 23°10'52" E 57°12'14".

EP 76

Wadi al Nakhr, "Grand Canyon of Arabia"
Note: You are standing on top of the Natih formation (Nt) and the thin underlying beds belong to the Nahr Umr formation (Nr).
Topics: Cretaceous and Jurassic sequence, spectacular scenic view
Location: UTM 40 Q 520870 2567095 / N 23°12'46" E 57°12'14"

Rating: ☺☺☺

Location map see EP 75, Fig. 171.

Approach: Follow road #21 from Nizwa towards the north-west and turn right after about 25 km into a minor road to Al Hamra. Turn left at the main roundabout in Al Hamra (UTM 40 Q 528410 2554703 / N 23°06'02" E 57°16'38") and follow the road for about 40 km. At the next 4 intersections always turn right (after 16, 24, 29, and 35 km). The viewpoint is at the top of the canyon.

Wadi al Nakhr is often referred to as the "Grand Canyon of Arabia" because it forms a narrow gorge with cliffs higher than 1,500 m in some places. It gets its name from the Arabic word Nakhr, which means 'incision'. From the eastern side, the wadi is flanked by Jebel Shams, the highest mountain peak in Oman with an elevation of more than 3,000 m. At the mouth of the wadi, the ruins of the old village of Ghul can be seen (EP 75). This has now been abandoned for a new village with modern houses on the other side of the wadi. The wadi continues for about 12 kms inside the southern flank of Jebel Akhdar. Recent travertine, forming stalactites and stalagmites, have been deposited along fault zones, cliff walls and porous rock units. Water flows at a number of places along the wadi, and after rain the water stream continuously runs for months and might make the road inaccessible.

A well-known flagged trekking path runs parallel to Wadi al Nakhr, from the village of Al Khateem at the ground level to the mountain top. This path is known as the balcony walk.

The spectacular gorge of the Wadi al Nakhr cuts through the entire Cretaceous and Jurassic carbonate sequences. It exposes the Sahtan, Kahmah and Wasia groups in great detail (Fig. 174). These sequences form cliffs more than 1,000 m high. The Sahtan group is Jurassic in age and consists of rust-brown shaley units overlain by bluish car-

Fig. 174: View into the canyon of Wadi al Nakhr. Indicated are the most important stratigraphic groups of Cretaceous and Jurassic rocks visible in the walls of the canyon.

bonates found in the deepest sections in the northeastern part of the Wadi al Nakhr. More common throughout the cliffs are the Wasia and Kahmah groups. The Kahmah group is a thick carbonate sequence which is interpreted as a mega-sequence showing a transition from deeper marine pelagic to shallow-marine limestones. The Wasia group unconformably overlies the Kahmah group. Evidence of sub-aerial exposure between the two groups can be seen in a number of places in Wadi al Nakhr. The Wasia group is composed of Nahr Umr shale and Natih limestone. The Nahr Umr formation forms a retreating slope due to its susceptibility to weathering. It mainly consists of brown marly *Orbitolina*-rich limestone. The Natih formation is divided into seven major units which can be described as alternation between competent shallow-marine limestone to incompetent, often organic-rich, argillaceous limestone units.

Various normal, reverse (close to the mouth of the wadi), strike-slip and oblique-slip faults can be observed, particularly in the outcrops of the Natih formation and Kahmah group. Numerous calcite-filled fractures developed in these once highly over-pressured rocks. The fractures form various patterns and have different relationships with the mechanical stratigraphy of the rocks. The geometry of faults and fractures can be studied in great details in Wadi al Nakhr. Most of these features trend WNW-ESE and show an open-mode component, in most cases cemented with white calcite that highly contrasts with dark grey colour of the carbonates. They were formed and reactivated during Late Cretaceous and Tertiary.

EP 77

Sint Megalodonts
Topic: Triassic limestone, megalodon shells
Location: UTM 40 Q 510614 2554483 / N 23°05'56" E 56°06'13"

Rating: ☺☺

For location map see EP 75, Fig. 171.

Approach: Follow the road from Al Hamra towards the north-west leading to the Jebel Akhdar. About 6.6 km after Wadi Ghul, turn left onto an unpaved road (at UTM 40 Q 515265 2562734 / N 23°10'24" E 57°08'56"). Drive down this road until you reach a paved road which leads to Sint and turn left there (at UTM 40 Q 507317 2561566 / N 23°09'46" E 57°04'17"). About 4 km after the village of Sint you will reach a parking lot at a left turn (UTM 40 Q 510594 2554622 / N 23°06'00" E 57°06'12"). Leave the car in the parking area and continue on foot towards the south. The outcrop starts about 200 m downhill.

The path that leads downhill to EP 78 and 79 from the parking area, follows the dip of a 1–3 m thick bed of a massive grey limestone. This limestone formed in a proximal intra- to supra-tidal depositional environment and makes up the upper part of the Late

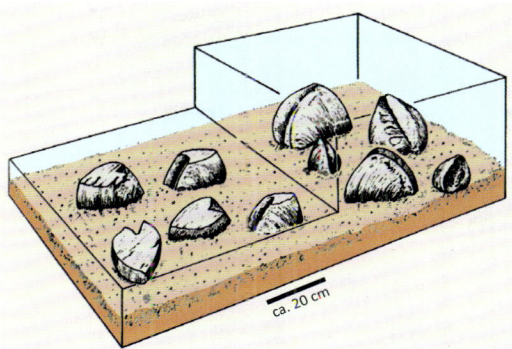

Fig. 175: The block diagram shows the animal's living position in carbonate mud (right) and the cutting view (left) on the rock surface through weathering (modified after Krumbiegel & Krumbiegel 1981).

Triassic Misfah formation. The rocks of this formation form Jebel Kawr and dominate the surrounding areas of the villages Sant and Sint. The surfaces are packed with thick-shelled bivalves in different cutting-views. These fossils are the most eye-catching feature of the rock (Fig. 175, 176). They belong to the extinct bivalve-family of Megalodontidae. The benthic living animals reached a considerable size of up to 20 cm and lived semi-infaunal as suspension feeders. The abundance of the Megalodon shells is very different in the layers of the Misfah formation. Some of them may be interpreted

Fig. 176: Megalodonts of different sizes on the stratum surface.

as subtidal storm deposits. Various cross sectional views of the shells can be studied here. In almost all cases the shells are still articulated, indicating their living positions.

EP 78

Sint Polje
Topic: Holocene lake depression
Location: UTM 40 Q 510775 2553564 / N 23°05'26" E 57°06'18"

Rating: ☺

For location map see EP 75, Fig. 171.

Approach: See EP 77 – continue for about 1 km downhill on the footpath to EP 77 into the flat depression.

The almost circular Holocene depression of Sint has a diameter of 500 m and is surrounded by karstified Late Triassic dolomite (Fig. 177). The plain is usually dry as it is part of the mountain desert environment. However, precipitation can be up to 200 mm per year, predominantly as rain during the winter months. The sediments filling the depression are washed in by surface runoff as sheet flows and only partly channelised flows from the adjacent mountains. The precipitation events are infrequent but heavy and wash off the debris resulting from physical weathering. Consequently, the depression has a very flat topography and is filled with sandy, silty and clayey material. The

Fig. 177: Holocene dry lake (polje) near Sint.

Fig. 178: Large trees (*Ziziphus spina-christi*) take advantage of the water resources available.

geomorphologic terminology for such a landform within a karstified environment is *polje*; a term that is derived from Slavic languages.

Artefacts in the form of thousands of flint chips and scattered flint tools can be found along the shores of the polje. These finds indicate that the site was used as a resting place for hunter communities at least during the winter months when migrating birds used the resources offered by the water-filled polje. Thankfully, even today, the large trees (Fig. 178) make the area a relaxing resting place in the otherwise barren landscape.

EP 79

Sint, Canyon Wadi al Alá
Topic: Canyon formation, conjugate fault structure
Location: UTM 40 Q 510110 2552908 / N 23°05'05" E 57°05'55"

Rating: ☺

For location map see EP 75, Fig. 171.

Approach: See EP 77 – continue 1 km downhill on the footpath to EP 77 into the flat depression (EP 78) and walk about 1 km towards the south-west to the edge of the canyon of Wadi al Alá.

The site described represents the southern face of the spectacular vertical incised Wadi al Alá (Fig. 179) which drains the Sint depression and marks the northern flank of Jebel Kawr. The canyon is 300 m deep. Exposed here are Late Triassic autochthonous shallow-water carbonates of the Misfah formation. Dominating lithologies include biolithoclastic limestone, dolomite as well as grey dolomitic limestone.

The deposits represent shallow-water platform relics that formed in a proximal subtidal and partially supra-tidal environment (Pillevuit et al. 1997). The rocks formed on an isolated carbonate platform representing various facies comprising platform-rim and bedded inner-platform facies with a total thickness of 800 m (Weidlich & Bernecker 2007). Jebel Kawr is regarded as one of the so-called "Oman exotics" (cf. Searle & Graham 1982) and is interpreted as an atoll-type seamount that formed in the Neotethys and is part of the Hawasina basin tectonic nappe. The interpretation is supported by the lithologies representing the base of the formation. Here, trachy-basalt and basalt are described, and the unit is seen as characteristic for sub-aerial volcanism (Beurrier et al. 1986). An outcrop of the severely weathered corresponding rocks can be found along the road towards Sint at UTM 40 Q 510079 2556222 / N 23°06'52" E 57°05'54".

Fig. 179: Canyon of Wadi al Alá.

The rock sequence is cut by several fault structures which require careful observation to be understood in detail. At the observation point listric normal faults, small-scale graben structures, and conjugate fault systems can be observed on the opposite wall of the canyon. The fault structures indicate extensional deformation.

EP 80

Al Hayl chevron folds
Topic: Hawasina nappe, chevron folds
Location: UTM 40 Q 502592 2567450 / N 23°13'00" E 57°01'30"

Rating: ☺

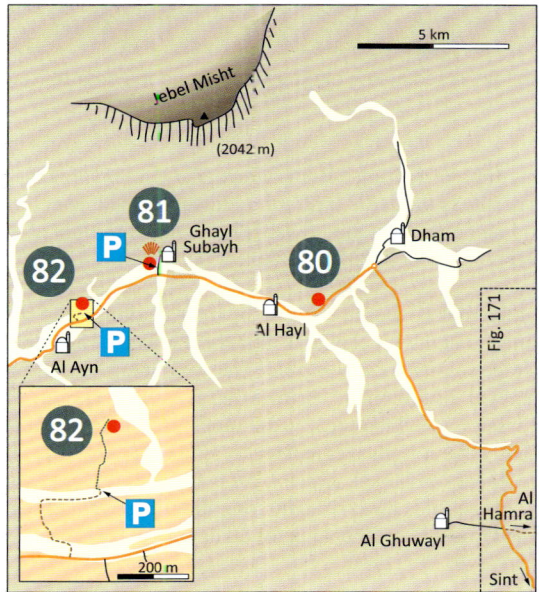

Fig. 180: Location map

Approach: Take the Wadi al Ayn road that connects with the Nizwa-Ibri road at the intersection close to UTM 40 Q 481551 2553169 / N 23°05'13" E 56°49'11" (under construction at the time of writing). Alternatively take the Amla-Wadi al Ayn road at the intersection close to UTM 40 Q 492363 2545636 / N 23°01'08" E 56°55'31" (under

Fig. 181: Chevron fold in sediments of the Hawasina nappe near Al Hayl.

construction at the time of writing). Both routes take you towards the Al Ayn beehive tombs (EP 82). The site is located 7 km further down the road from there. You can also take the road winding down from Sint towards Barut and turn right at the roundabout at UTM 40 Q 503916 2568494 / N 23°13'32" E 57°02'17".

The lithologies of the rocks exposed close to the road are dominated by decimeter thick limestone beds intercalated with clay and siltstone. Graded lamination with small current ripples and trace fossils are observable. The rocks represent turbiditic sequences, deposited in a bathyal environment. Beurrier et al. (1986) assign a Late Triassic to Early Jurassic (Liassic) age. Tectonically, the rocks form part of the Hawasina nappe. The limestone beds have a higher competence in comparison with the clay and siltstone. Symmetric, V- or M-shaped, *chevron folds* formed as a result of compressional forces (Fig. 181). These folds have straight, long limbs and a sharp, narrow hinge. The folds here have the typical interlimb angle of 60 degrees.

EP 81

Jebel Misht viewpoint
Note: The point given here is not on top of the mountain but a viewpoint which is an excellent stop for a lunch break under one of the large trees in the valley
Topic: Oman exotics
Location: UTM 40 Q 498231 2568389 / N 23°13'28" E 57°58'57"

Rating: ☺☺

For location map see EP 80, Fig 180.

Approach: Take the Wadi al Ayn road that connects with the Nizwa-Ibri road at the intersection close to UTM 40 Q 481551 2553169 / N 23°05'13" E 56°49'11" (under construction at the time of writing). Alternatively take the Amla-Wadi al Ayn road at the intersection close to UTM 40 Q 492363 2545636 / N 23°01'08" E 56°55'31" (under construction at the time of writing). Both routes take you towards the Al Ayn beehive tombs (EP 82). The site is located 2.5 km further down the road from there. You can also take the road winding down from Sint towards Barut and turn right at the roundabout at UTM 40 Q 503916 2568494 / N 23°13'32" E 57°02'17".

Jebel Misht is one of the so-called "Oman exotics". This term was coined by Glennie et al. (1973, 1974) in their initial description of Oman's geology. The exotics are usually isolated mountains made up of limestone and dolomite (Searle & Graham 1982). They appear exotic, as they are allochthonous and commonly associated with darker sediments and rocks from the ophiolite sequence from which they stand out by their

Fig. 182: View of the Oman exotic block of Jebel Misht. Interpretations of the stratigraphic succession after Searle & Graham (1982).

lighter colour. Jebel Misht is the most spectacular Oman exotic in terms of geomorphology, as the mountain is located in dramatic scenery. Therefore, Searle (2014) proposed Jebel Misht to be included in the list of geological heritage sites of Oman that could be part of a geopark. The mountain appears like a cliff, surrounded by Hawasina sediments. The mountain represents a cuesta which is an asymmetric ridge with an escarpment on one side and dip slope on the other side. The almost vertical rock face is 1,000 m high and visible from Jebel Shams. The Arabic translates to "comb mountain" reflecting the fact that the top of the mountain appears highly irregularly weathered.

According to Searle & Graham (1982) the Late Triassic shallow water carbonate rocks (reefal limestone) in the upper part of the profile represent the highest thrust slice in the allochthonous Hawasina complex. They are thrust over younger Hawasina sediments and alkali volcanic rocks in the form of black, vesicular pillow lavas and microdolerite (Fig. 182). The general interpretation is the formation of the carbonates on top of a seamount, also known as guyot that formed several kilometers away from the Arabian continental margin (Searle 2014). Red and greenish radiolarian chert, lithoclastic or micritic limestone and quartzose sandstone are described by Beurrier et al. (1986) at the foot of Jebel Misht. Several listric normal faults are observable within the rock face (see Fig 182).

EP 82

Al Ayn beehive tombs
Topic: Beehive tombs, sediments Hawasina nappe
Location: UTM 40 Q 496140, 2567436 / N 23°12'57" E 56°57'44"

Rating: ☺☺☺

For location map see EP 80, Fig. 180.

Approach: Take the road from Sint to Al Hayl. Turn left at the roundabout in Dham and continue on the road passing through Al Hayl. 6 km after Al Hayl, the beehive tombs of Al Ayn are on the right hand side of the valley well visible from the street. A small unpaved road through the village leads to a parking area in front of the archeological site (400 m).

The beehive tombs of Al Ayn (Fig. 183) are listed as a UNESCO World Cultural Heritage (UNESCO 2014a). The chain of tombs is located in impressive scenery on the crest of a mountain range in the shadow of the impressive Jebel Misht (EP 81); they overlook the wadi and the village of Al Ayn. The 21 tombs of the Hafit type (see EP 50, Halban) are remarkably well preserved. Together with the necropolis of Bat which is 25 km to the north-west and the tower of Al Khutm which is 2 km west of Bat, the tombs are protected as the most complete and best-known site of the early Bronze Age (3rd mil-

Fig. 183: The Early Bronze Age beehive tombs of Al Ayn in front of the Jebel Misht massive.

lennium BCE). The three locations reflect the emergence of a strictly structured hierarchic society. This development was linked to the copper trade.

During the 3rd millennium BCE, the ancient country of Magan, a region which corresponds to Oman, represented the centre of copper extraction. From here, copper was traded as far as to Mesopotamia. The trade brought higher living standards and social changes visible in the changed structure of settlements and the array of funerary space. Al Ayn has not been archaeologically excavated until now.

The tombs of Al Ayn are two-walled and assembled of dressed limestone that crops out at the location (Fig. 184). Due to weathering, the rock slabs appear yellowish; fresh cutting surfaces reveal a grey colour. The bedded limestone shows sedimentary features

Fig. 184: Beehive tombs of Al Ayn.

like ripples and sole marks. The beds partly contain Bouma sequences. These turbidite deposits accumulated in a bathyal environment during the Jurassic/Cretaceous period. The deposits are mapped as Sid'r Formation ranging from Late Tithonian to Early/Middle Cenomanian (Janjou et al. 1986).

EP 83

Wadi Murri
Topic: Structural geology, plunging fold
Location: UTM 40 Q 504779 2592407 / N 23°26'29" E 57°02'48"

Rating: ☺☺

Fig. 185: Location map.

Approach: Follow road #10 from Rustaq towards Ibri. Leave the main road at the intersection at UTM 40 Q 502974 2596671 / N 23°28'48" E 57°01'44", follow the road for 5.6 km passing through the small village of Murri. About 2.5 km after the village, turn left into a narrow wadi. Leave the car at the exit of the wadi and start walking into a small but impressive gorge. There is a second but smaller section through the sequence which you passed on your way at UTM 40 Q 503918 2594496 / N 23°27'37" E 57°02'18". If you continue on the road towards the south, you will eventually end up at Jebel Misht.

The road is located in a wadi that divides the allochtonous rocks of the Samail Ophiolithe to the west from autochthonous rocks of the Al Hajar Mountains to the east. There are some small remnants of rock formations belonging to the Hawasina nappe in between. As the base of the ophiolite sequence is made up of a thrust, you may also encounter discontinuous outcrops of the metamorphic sheet along the road.

The autochthonous rocks to the east form a large anticlinal structure predominantly made up of dark limestone with the fold axis plunging towards the north-northwest. They belong to Early and Late Cretaceous formations (upper formation within the Kahmah group; Nahr Umr and Natih formations; cf. Fig. 172 of EP 75). Commonly massive, partly weakly laminated limestone alternating with calcareous siltstone is exposed (Beurrier et al. 1986). Observable within the formations are rudists and large foraminifera. These formations formed along the passive margin of the African Plate (Arabian margin).

The rivers cutting through the mountains, flowing against the prevailing topography are referred to as antecedent streams. This is clear evidence that the drainage pattern must have developed before the uplift of the underlying strata. Hence the geomorphology indicates that the topographic framework of the Al Hajar Mountains must be very young and that mountain building is ongoing.

The massive beds of limestone are polished by fluvial activity in the lower parts of the valleys. These canvases are decorated with rock art in the form of petroglyphs. Human figures, often with weapons and animals, predominantly camels, are depicted. They presumably represent hunting scenes.

EP 84

Wadi al Hibi (southern part) sheeted dikes
Topic: Sheeted dikes, Samail Ophiolite
Location: from   UTM 40 Q 452256 2660554 / N 24°03'24" E 56°31'50"
          to   UTM 40 Q 452256 2660554 / N 24°03'02" E 56°31'29"

Rating: ☺☺

Approach: Take the road #8 from Yanqul to Sohar. About 1 km before Hayl al Ashqarain, take a newly constructed road towards the west (at UTM 40 Q 454654 2660254 / N 24°03'13" E 56°33'13") and follow this road for 2.7 km to Wadi al Hibi. Take a sharp

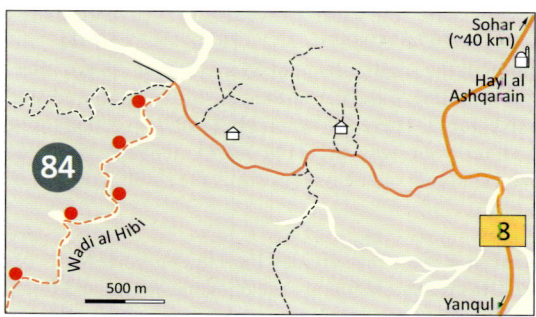

Fig. 186: Location map.

left in the little village entering into the wadi. The outcrop starts about 300 m after entering the wadi.
The sheeted dike complex is a unit of more of less vertically intruded dikes of the oceanic crust formed at a mid-oceanic spreading center below the pillow basalts and above the plutonic unit with mostly gabbroic rocks (Fig. 18). Since their appearance resembles that of a layered rock, they are called a sheeted dike complex, although this is not a layering in the classical sense. They are considered to have formed at an active mid

Fig. 187: Sheeted dike complex in the Wadi al Hibi.

Field sites

Fig. 188: Detail of the sheeted dike complex with a chilled margin of a half-dike.

oceanic spreading center during repeated intrusions of basaltic melt into fractures (Fig. 20; Frisch et al. 2011, Frisch & Meschede 2013). Generally, they are around 50 cm to 1 m thick and intrude into one another. A dike is characterised by finer grained edges since the edges cool faster forming the so-called chilled margins. In the center of the dike the crystals have more time to grow and dolerites can develop. A new generation of dikes may intrude into the older ones which are not yet solidified completely and, therefore, easy to split in two halves. These half-dikes are fine-grained on one side and coarser grained on the other, which represents the former center of the older dike. Structures of this behavior are called dike-in-dike structures and are indicative for the formation at an oceanic spreading center.

An extraordinary well exposed series of sheeted dikes can be studied at several places in the Wadi al Hibi at the northeastern flank of the Central Oman Mountains (Jebel Moqalit/Jebel Aswad). Nearly vertical dikes of mostly 0.5 to 2 m thickness are exposed along the walls of the wadi (Fig. 187), showing dikes and half-dikes (Fig. 188) as well.

EP 85

Lava tube Wadi al Hibi (northern part)
Topic: Basaltic extrusion, Samail Ophiolite
Location: UTM 40 Q 454283 2666442 / N 24°06'34" E 56°33'00"

Rating: ☺☺

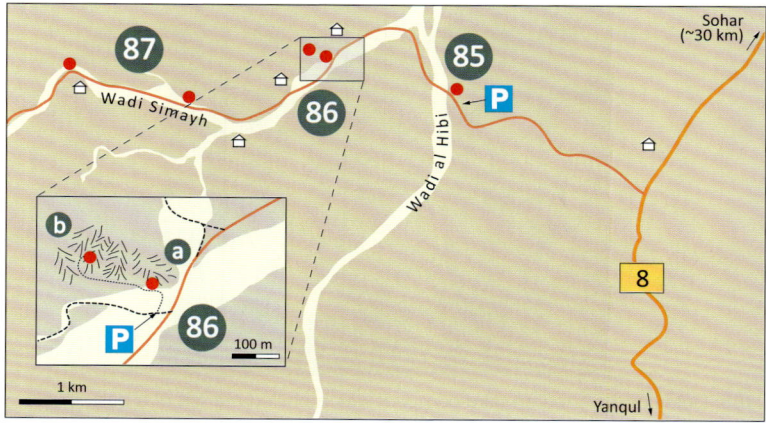

Fig. 189: Location map.

Approach: Take the road #8 from Yanqul to Sohar. At UTM 40 Q 456202 2665474 N 24°06'03" E 56°34'07" turn left and follow the road for about 2.4 km. The outcrop is directly beside the road.

At this outcrop a very nicely exposed lava tube (Fig. 190) that was formed during the extrusion of the basaltic lavas of the Samail Ophiolite can be observed. The tube has a diameter of about 5 m. The basalt shows chilled margins at the outer rim of the layers, indicating a rapid cooling at the contact with the surrounding material. The basaltic or andesitic lavas belong to the upper extrusive series of the late magmatic unit (Le Métour et al. 1992).

A sill beside the lava tube reveals columnar structures of up to 30 cm in diameter. An alteration horizon at the base of this sill was formed by ocean floor metamorphism. This process called spilitisation releases calcium (Ca) from plagioclase and replaces it with sodium (Na) from the seawater. The plagioclase becomes more and more albitised and pyroxene and olivine are chloritised. Moreover, the released Ca causes the growth of epidote which gives the rock its characteristic grass-green colour (Fig. 191). Vesicular cavities indicate an extrusion of the lava at a depth of less than 1500 m.

Fig. 190: Lava tube in the northern part of Wadi al Hibi.

Fig. 191: Chilled margin at the base of a lava flow. The grass-green colour is caused by epidote.

EP 86

Wadi Simayh, columnar basalt
Topic: Basalt of the Samail Ophiolite, columnar structures
Location: a) UTM 40 R 452919 2666837 / N 24°06'47" E 56°32'12"
b) UTM 40 R 452791 2666892 / N 24°06'49" E 56°32'07"

Rating: ☺☺

For location map see EP 85, Fig. 189.

Approach: Leave the main road #1 from Muscat to Sohar in Sohar at the first large roundabout, taking the exit towards the southwest to Yanqul. Follow road #8 for about 28 km up to the junction with the Wadi al Hibi road. Turn right into the Wadi al Hibi road and follow it for about 4.1 km on the paved road (1.5 km west of EP 85). The outcrop appears on the right hand/ northern side of the road.

Columnar basalts occur at this outcrop in a striking way. The lava formed as thick layers during the formation of the Samail Ophiolite. They belong to the lower extrusive series of the early magmatic sequence (Le Métour et al. 1992). Diameters of the basaltic columns are between 20 cm up to half a meter. Horizontal as well as vertical cuts of the

Fig. 192: Horizontal section through columnar basaltic lava at location EP 86-a (see location map at EP 85) in the Wadi Simayh within the lower extrusive sequence of the Samail Ophiolite.

Fig. 193: Vertical columnar basalt at location EP 86-b (see Fig. 189) showing the stepwise cooling of the basaltic layer. The horizontal layering is well visible because of the weathering of the rocks.

columnar basalts are exposed and show the predominantly hexagonal structures (Fig. 192). Vertical sections indicate the stepwise development of the columnar joints. A horizontal layering of 5 to 10 cm is clearly visible (Fig. 193). This is a result of the stepwise propagation of a fracture in the transition zone between the already cooled, brittle reacting part and the still-ductile reacting hot part of the lava flow. With the proceeding of the cooling a tensional force develops between those two parts which episodically leads to the propagation of the tensional fractures that form the columnar structures.

EP 87

Wadi Simayh, sheeted dikes
Topic: Sheeted dikes, Samail Ophiolite
Locations: UTM 40 R 451586 2666353 / N 24°06'31" E 56°31'25"
          UTM 40 R 450445 2666704 / N 24°06'42" E 56°30'44"

Rating: ☺☺

For location map see EP 85, Fig. 189.

Approach: Leave the main road #1 from Muscat to Sohar in Sohar at the first large roundabout, taking the exit towards the southwest to Yanqul. Follow road #8 for about 28 km to the junction with the Wadi al Hibi road. Turn right onto the Wadi al Hibi road

Fig. 194: Slightly tilted sheeted dikes of the early magmatic unit of the Samail Ophiolite.

and follow it for about 5.8 km (location a) and then to 7.2 km (location b) on the paved road (further west of EP 84 and EP 85). Both outcrops appear on the right hand/northern side of the road.

In this outcrop a series of sheeted dikes is exposed that is slightly tilted towards west (Fig. 194). Originally they were formed in an upright position when the dikes intruded into the older ones. The tilting is a result of tectonic deformation during the nappe movement of the Samail Ophiolite thrust over the sediments of the Hawasina unit. For further explanation of sheeted dike structures see EP 84.

EP 88

Sahban spring in Wadi al Jizzi
Note: A 4-wheel vehicle is highly recommended for the graded roads leading to this outcrop.
Topic: Alteration of ultramafic rocks, calcite precipitation
Location: UTM 40 R 430520 2675012 / N 24°11'10" E 56°18'57"

Rating: ☺☺

Approach: Follow the express road #7 from Sohar to Al Buraimi for 37 km. Take the exit to the village of Sahban (at UTM 40 R 431780 2679514 / N 24°12'35" E 56°19'33") and pass under the bridge towards the south. After about 400 m, turn right and follow the graded road towards the west and then south for 4.3 km. At UTM 40 R 430952 2676076 / 24°11'45" E 56°19'12" turn right and follow the graded road for another

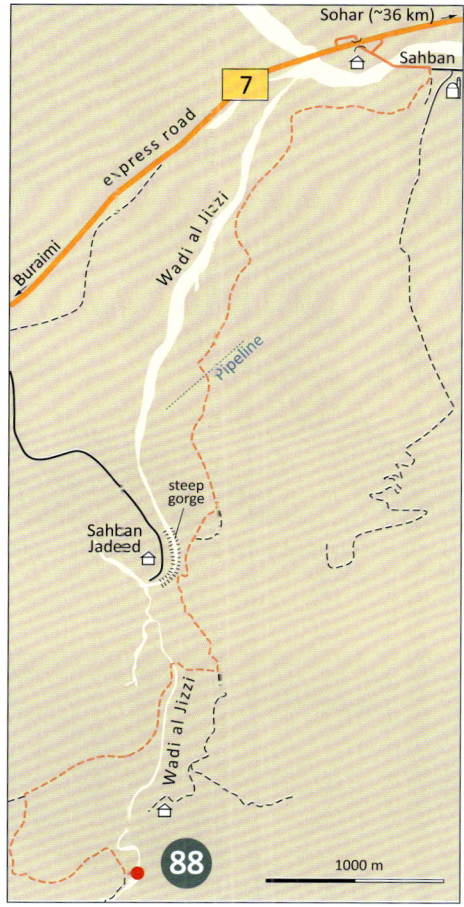

Fig. 195: Location map.

1.4 km, then turn left. After about 1.2 km you will notice the white water pool of Sahban on your left hand side.

The natural white water springs of Sahban run continuously throughout the year with an almost constant flow of water. They are well known in the area for the beautiful white and blue colours that form deposits of white fine travertine (Figs. 196, 197). This process of fresh water mineral precipitation is considered to be a possible solution for the

Fig. 196: White pools in serpentinised ultramafic rocks of the Samail Ophiolite with recent calcite precipitation in the Wadi al Jizzi.

Fig. 197: Sheets of solid carbonatic minerals that precipitated on the surface cover the bottom of the pools.

reduction of the increasing concentration of CO_2 in the atmosphere (Kelemen & Matter 2008, Matter & Kelemen 2009). The precipitation of solid carbonatic minerals results from combining CO_2 with divalent cations such as Ca^{2+}, Mg^{2+}, and Fe^{2+}. The most abundant geologic sources of these cations are silicate minerals such as olivine, pyroxene, serpentine and plagioclase. These are primarily common in basaltic and peridotitic rocks as they are found in the Samail Ophiolite sequence which occurs around Wadi al Jizzi. Peridotite, an igneous rock derived from the Earth's mantle contains the minerals olivine $((Mg,Fe)_2SiO_4)$ and pyroxene $((Ca,Mg,Fe)_2Si_2O_6)$. Both minerals have high concentrations of magnesium, iron and calcium.

At the Earth's surface mantle rocks are significantly out of equilibrium and, therefore, react with water and CO_2 at a high rate compared to other rocks. In the presence of water peridotite is serpentinised. The mineral serpentine $(Mg_3Si_2O_5(OH)_4)$ reacts with CO_2 and turns into talc $(Mg_3Si_4O_{10}(OH)_2)$ and magnesite $(MgCO_3)$. Both minerals are abundantly present as veins in the mantle sequence of the Samail Ophiolite (e.g., EP 01, EP 60). When pyroxene is involved in the reaction as well, the product also includes calcite $(CaCO_3)$ (cf. EP 57). There are several factors affecting this reaction, such as pressure, temperature, salinity and water acidity (Dabirian et al. 2012). Recently performed studies show that millions of tons of CO_2 could be absorbed by peridotite if CO_2 would be artificially injected into deep vertical wells (Kelemen et al. 2011).

EP 89

Wadi al Jizzi pillow basalts
Note: A field trip to Oman is not complete without a visit to the world renowned, extraordinary well exposed outcrops of pillow basalts in the Wadi al Jizzi.
Topic: Pillow basalt formation, dike-in-dike structure
Location: UTM 40 Q 438300 2685899 / N 24°17'05" E 56°23'31"

Rating: ☺☺☺☺

Approach: From the Globe roundabout in Sohar, drive 25 km in a southwesterly direction to Al Wasit on the newly constructed main road #7, the main road from Sohar to Al Buraimi. Leave the 4-lane highway at the exit Falaj al Qabail (at UTM 40 Q 441060 2686369 / N 24°17'21" E 56°25'10"), then turn right onto the Suhayjah road and follow the old dual carriageway #7 into Wadi al Jizzi. Continue for about 3.5 km on the old Wadi al Jizzi road. The outcrop of the so-called "Geotimes pillow lava" is on the left hand side of the road, well visible from the bridge crossing Wadi al Jizzi. There is a gravel track with an entrance at UTM 40 Q 438172 2685792 / N 24°17'02" E 56°23'26") at the southwestern side of the wadi bridge next to the outcrop. This gravel road leads down into the valley and to the outcrop.

Wadi al Jizzi is famous for its magnificent outcrops of the Samail Ophiolite. The pillow lavas of the so-called "Geotimes outcrop" belong to the older series (v1) of submarine

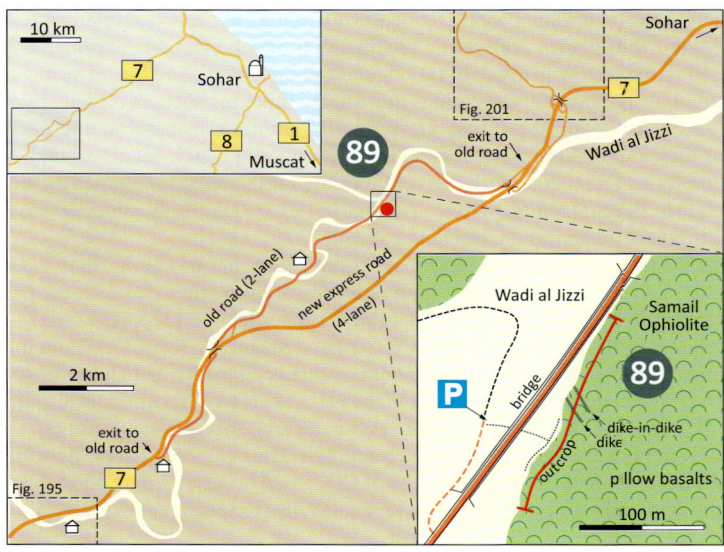

Fig. 198: Location map.

volcanic rocks of the Samail Ophiolite. This outcrop has extraordinarily fine quality examples of the pillow structures formed at an oceanic spreading center at the seafloor. The outcrop is unique and well known as one of the best outcrops of pillow lavas worldwide. The term "Geotimes pillow lava" was first used by an international field conference on ophiolites in 1972 (Anonymous 1972) that published an article about the pillow lavas of Wadi al Jizzi. The extraordinary good preservation of the pillows is probably due to the fact that they have been covered with alluvium until recent times.

Numerous large pillow basalts of 0.5 to 1 m in diameter and lava tubes more than 2 m in length occur on a surface which is tilted at about 50–60° towards the northwest, representing the ancient ocean floor (Fig. 199). The pillows cover a surface of at least 50 m high and 100 m long. In some pillow structures cm-scale cooling fractures with small columnar structures can be observed. Triangular spandrels between the pillows contain hyaloclastic or sedimentary material.

Pillow basalts form at submarine outflows of lava when hot molten lava extrudes into cold seawater where it is immediately cooled down forming a chilled margin (Fig. 20). This chilled margin is the skin of the pillows which consists of basaltic glass where the glass is a product of the sudden cooling by cold sea water. This process occurs so fast that crystallisation of single crystals is not possible.

Molten rock material which suddenly gets in contact with cold water ideally cools in a spherical form because a sphere provides the smallest surface per volume unit thus

Fig. 199: Pillow lavas of the "Geotimes" outcrop in Wadi al Jizzi.

reducing the heat loss. However, the lava pillows solidify in a slightly oval form because of flowage of lava outward and downward due to the weight of the pillows. Moreover, the underlying already cooled pillows form an irregular concave surface which is filled up by the new pillows. The spherical bodies remain liquid for a while and have a primarily thin and still formable skin of basaltic glass. Therefore, they are able to fill up the irregular surface. The irregular triangular spandrels indicate the original orientation of the pillow layer and may serve as a fossil horizontal marker. If a large amount of lava erupts, the lava will flow in tube-like structures until they end in spherical objects. These tubes can also be observed in several variations in the outcrop.

Basalts primarily evolve as a mixture of mainly pyroxene and plagioclase. The chilled margins of the pillow consist of glass representing the composition of the melt. Glass is a general term for non-crystalline, amorphous rock material and the atoms are not arranged in a crystal lattice but are distributed in a disorderly way such as in a pane of window glass. Basaltic glass is often transferred into chlorite over time and with slight heating. Plagioclase which originally contains 50–55 % Ca-plagioclase (anorthite) and 45–50 % Na-plagioclase (albite) is rapidly metamorphosed to albite when the Ca-components are used to produce crystals of epidote, partly as inclusions in plagioclase. These "filled" or "saussuritised" plagioclases have a light green colour. The term "saussuritisation" is used in honor of the Swiss natural scientist Horace-Bénédict de Saussure who lived in the 18th century (1740–1799).

Fig. 200: Dike-in-dike structure with chilled margins crosscutting the pillow basalts of the 'Geotimes' outcrop of Wadi al Jizzi. For location of the dikes see sketch map of the outcrop.

Two dikes crosscut the Geotimes outcrop; here the internal structure of dikes can be observed. The northern dike represents a dike-in-dike structure where a younger dike (dike 2) intruded into a slightly older one (dike 1; Fig. 200). It is assumed that the older dike was still hot in its inner part when the second dike intruded. The second dike parted the older one into two halves forming two so-called "half-dikes" with only one chilled margin.

EP 90

Arja copper mine
Topic: Copper mining, VMS's, upper extrusives, Samail Ophiolite
Location: UTM 40 Q 440448 2692832 / N 24°20'51" E 56°24'46"

Rating: ☺☺

Approach: Follow the expressway #7 from Sohar to Al Buraimi for 24 km. Take the exit directly at the copper-smelting plant (at UTM 40 Q 441817 2687936 / N 24°18'12" E 56°25'35") and follow the paved road for about 7 km to EP 90a and then a further 700 m to EP 90b.

Wadi al Jizzi is one of the main east-west orientated wadis that cross the northern Oman Mountains. An ancient route existed between Sohar and the oases of Al Buraimi that ran through it. The area is famous for its copper ore and has been mined since at least 4,000 years ago (Weisgerber 1987). It is also known as the area that has been identified as the legendary country of Magan.

The Land of Magan is known as a copper source region for Mesopotamia from Sumerian cuneiform texts from Ur of around 4,300 years ago. Over 32 ancient copper smelting sites (Fig. 202) have been identified in the immediate area, and open-cast mines, abandoned only decades ago, exist in Lusail, Arja and Baida. Mining continues up to the present. There is a large operational copper-smelting plant around 25 km from Sohar, along the brand-new Sohar – Buraimi highway. In the early years of copper extraction, the ore was principally exported to the center of civilisation at that time; the Mesopotamian kingdoms such as the Chaldeans of Ur. Their religious expressions were centered on stepped pyramids or ziggurats, the remains of which can still be visited in present day Iraq and Iran. The earliest pyramids in Egypt were also ziggurats, but the Egyptians soon learned how to clad their pyramids resulting in the well-known smooth-sided pyramids such as in Giza.

There is only one ziggurat in the whole of the Arabian Peninsula (Costa & Wilkinson 1987). It is situated in Arja, very near to the ancient copper mines of Arja and Baida (EP 90b). Apparently when the first pre-historic mining engineers were sent by their Mesopotamian masters to develop copper mines in Oman, they took their favorite design for a place of worship along with them and constructed a miniature ziggurat next to their work place. The ziggurat in Arja consists of two stepped platforms and has a clearly recognisable processional ramp aligned with what was believed at that time to be the seat of the Gods, a conspicuous black mountain visible on the other side of the ziggurat. Ruins of various accompanying buildings can also be recognised next to it, when standing on top of the ziggurat. However, the interpretation as a ziggurat is questioned by Yule (2008).

The outcrop at EP 90a is part of the Arja copper mine complex near Wadi al Jizzi. It is an old and abandoned mine, where copper occurred as a massive sulfide deposit.

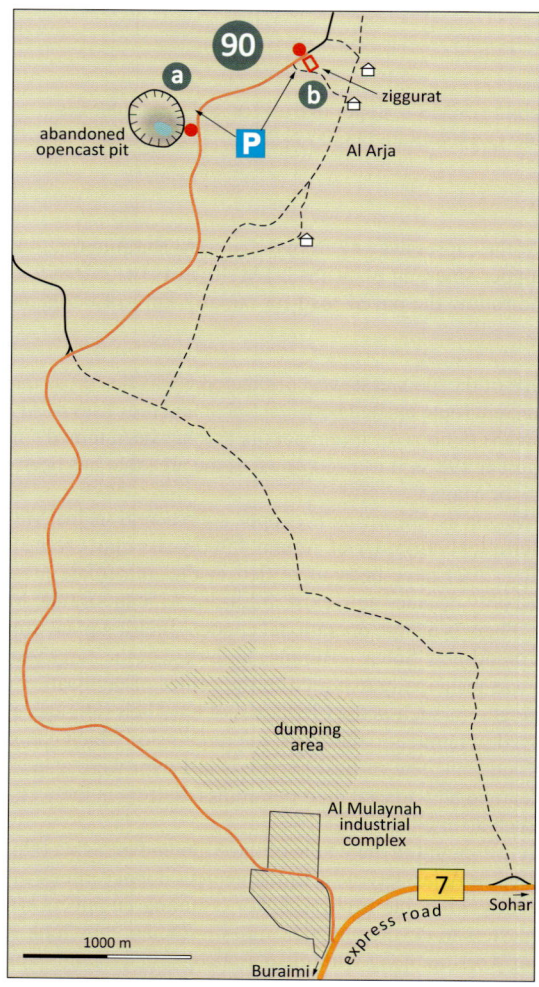

Fig. 201: Location map.

Field sites

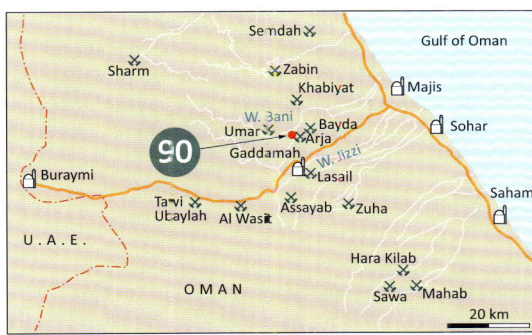

Fig. 202: Distribution of ancient copper mines in the vicinity of Wadi al Jizzi (modified after Weisgerber 1987).

Besides the iron ores magnetite and pyrite, most copper ores are composed of chalcopyrite ($CuFeS_2$) and some other sulfide copper ores. Secondary minerals are mostly chrysocoll and malachite. The copper ores were formed as volcanic hosted massive sulfides-deposits (VMS). Such deposits are currently formed at mid-oceanic spreading centers at black smokers. Hydrothermal fluids transport metallic ions which precipitate at the contact with cold seawater and form the deposits around the black smoker. The ore deposit consists of a massive sulfide ore body in the upper part and a stockwerk zone in the lower part, which represents the former feed channel level of the hydrothermal fluids. Generally, a gossan with secondary minerals like malachite and limonite is developed.

A visit to the copper mine other than enjoying the view into the digging hole from above is not possible because of safety reasons. The upper part of the ore body is char-

Fig. 203: The abandoned opencast pit copper mine of Arja.

acterised by a gossan with light yellow to brown colours (Fig. 203) indicating the enrichment of iron minerals. Beneath this zone follows the oxidation zone with the strongest enrichment of copper minerals up to the occurrence of solid copper. The real ore body follows below with copper minerals like chalcopyrite or bornite. The ore body dips obliquely, causing the oval shape of the open pit mine. Generally, the content of copper is below 1 %.

EP 91

Black Smoker southwest of Sohar (also known as Zuha gossan)
Topic: Hydrothermal activity, VMS formation, upper extrusives, Samail Ophiolite
Location: UTM 40 Q 452197 2675691 / N 24°11'36" E 56°31'45"

Rating: ☺☺☺

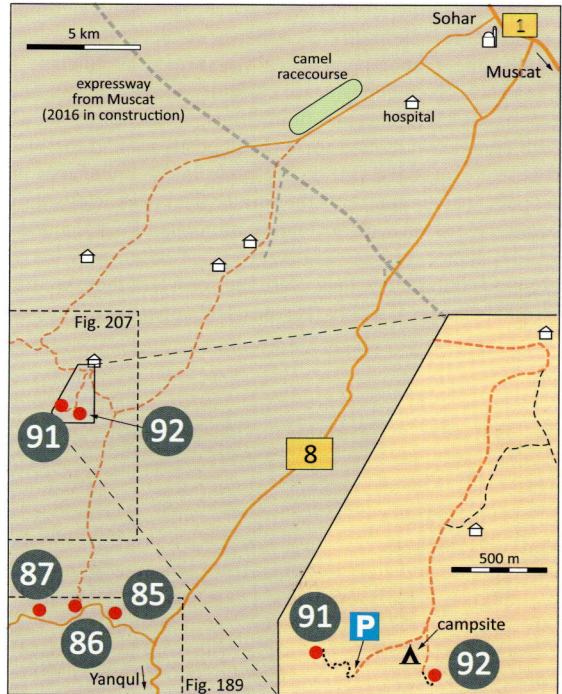

Fig. 204: Location map.

Field sites

Fig. 205: Red coloured rocks of the black smoker, Zuha gossan, southwest of Sohar, view from the southwest.

Approach: In Sohar, take the road to the hospital and continue straight, passing the camel racecourse. At the end of the camel racecourse, it is possible to take both roads. Follow the unpaved road to the left for about 15.5 km crossing the construction site of the expressway from Muscat, then turn right and continue on the road for another 1.8 km. Turn left (at UTM 40 Q 453473 2676945 / N 24°12'16" E 56°32'30") and follow a small unpaved road towards the southwest. The outcrop with red rocks on top of a hill is clearly visible from here.

The gossan is located at the stratigraphic top of the lower lava unit within the ophiolite sequence. The outcrop of the Zuha gossan is located on a fairly well visible prominent hill which catches the attention because of its unusual colours (Fig. 205). The overall red colour of the hill, and red and brown oxidised material are widely visible within the uniform dark brown and brown mass of basaltic rocks in the surroundings. Near the top of the hill, mostly white and greenish altered basalts contain numerous veins with secondary copper minerals such as malachite and azurite. Other copper minerals at this location are chalcopyrite, bornite, cuprite and pyromorphite (see Karpof et al. 1988, for a full geochemical characterisation of the deposits). The copper material has been explored for mining and mining of the deposit could start any time soon. Drilling campaigns revealed an ore body of 20 meters thick and the main gossan area covering 250 × 100 m. Copper mineralisation of 2.8 % has been proven.

The rock association of this hill is interpreted as a fossil "black smoker". It represents a vent structure on the ocean floor where hot fluids of superheated water loaded with dissolved sulfides, partly high concentrations of metals like copper, zinc and manganese extrudes into the cold seawater (Fig. 205). The contact with the cold water leads to the precipitation of minerals that in some cases form black, roughly cylindrical chimney-like structures around the vents of up to 60 meters in height. In Oman, the deposited metal sulfides became volcanic hosted massive sulfide-deposits (VMS) which are still mined at several locations.

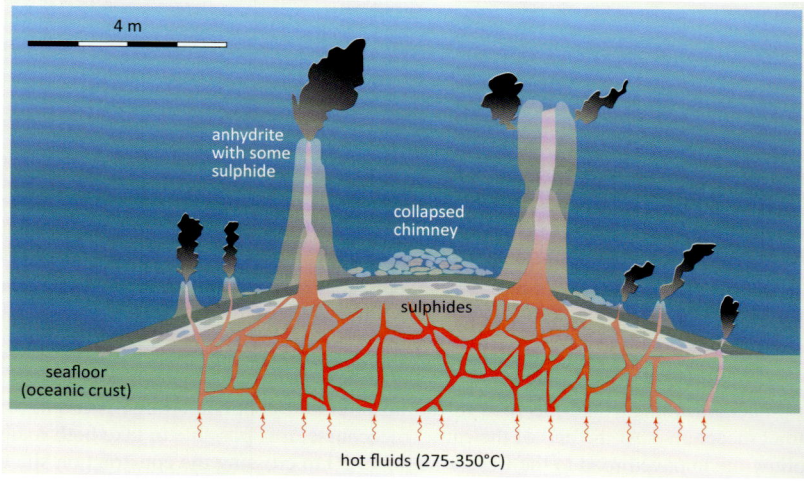

Fig. 206: Principle sketch of the function of black smoker and massive sulphide ore formation (VMS, modified after Evans 1993).

EP 92

Metalliferous sediments southwest of Sohar
Topic: Sediments of the Samail Opholite unit, metalliferous sediments
Location: UTM 40 Q 452831 2675680 / N 24°11'35" E 56°32'07"

Rating: ☺

For location map see EP 91, Fig. 204.

Approach: In Sohar, take the road to the hospital and continue straight, passing the camel racecourse. At the end of the camel racecourse, it is possible to take both roads. Follow the unpaved road to the left for about 15.5 km, then turn right and continue on the road for another 1.8 km. Turn left (at UTM 40 R 453473 2676945 / N 24°12'16" E 56°32'30") and follow a small unpaved road towards the south. Several outcrops with dark layers can be observed in the sediments.

The site is located in the direct vicinity to the Zuha gossan (EP 91) which is located around 600 m to the west (Fig. 207). The rocks of interest crop out along the flanks and the crest of a small hill. Here the unmetamorphosed sedimentary cover above the extrusive successions of the Late Cretaceous Oman ophiolite is exposed (Fig. 208). Pillow

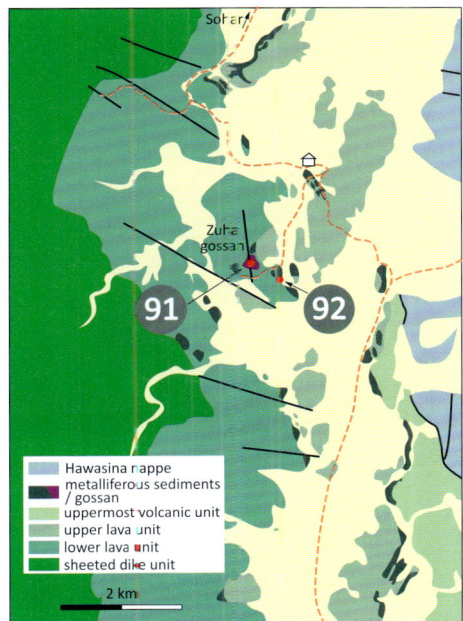

Fig. 207: Distribution of metalliferous sediments in the vicinity of the Zuha gossan. For location see EP 91, Fig. 204. Geological map modified after Karpoff et al. (1988).

Fig. 208: Layer of metalliferous sediments (dark grey) above pillow basalts of the uppermost units of the Samail Ophiolite.

Fig. 209: Initial stage of kink band structures in fine-laminated layers of the metalliferous sediments.

basalts are obvious especially on the eastern side of the hill. These extrusive rocks are overlain by fine-grained and parallel-laminated brown umbers, rich in iron and manganese and associated with radiolarian mudstones. Some of the fine-laminated layers show kink band deformation structures in an initial stage (Fig. 209). Hyaloclastic components can be found in the spaces between individual pillows.

The pelagic and metalliferous (siliceous, ferruginous and ferro-manganiferous) sediments formed in the vicinity of a marginal ocean-basin spreading axis (Wilson 1997). The Fe, Mn and trace metal-enriched sediments can be explained as precipitates formed by oxidation of hydrothermal inputs derived from high-temperature sulphide-precipitating vents (Robertson & Fleet 1986). Such systems are also known as black smokers (cf. EP 91). The associated hydrothermal processes in the deep sea are complicated and include leaching of volcanic rocks by hydrothermal fluids, Calcium metasomatism and the alteration of the primary phases by oxidation (Karpoff et al. 1988). Furthermore, the interaction of hydrothermal fluids and pelagic sediments are discussed (Fleet & Robertson 1980).

EP 93

White Smoker Al Ghizayn
Topic: Hydrothermal activity, VMS formation, upper extrusives, Samail Ophiolite
Location: UTM 40 Q 498838 2634848 / N 23°49'29" E 56°59'18"

Rating: ☺☺☺

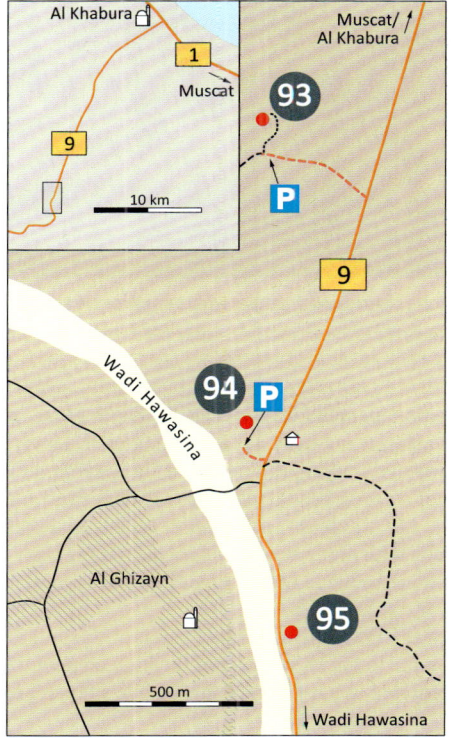

Fig. 210: Location map.

Approach: Take the road #9 from Al Khabura (road #1) to Al Ghizayn. After about 20 km, turn right into an unpaved road. After 400m, leave the car by the roadside and walk 100 m up the hill north of the road.

Fig. 211: Colourful rocks associated to the white smoker near Al Ghizayn.

The outcrop is located on a small hill and well visible because of its unusual colours. White altered basalts and red and brown oxidised material are visible within a uniform dark brown and brown mass of basaltic rocks in the surroundings. Various rocks are exposed near the top of the hill; brown, dark brown, dark grey and reddish chert layers (Fig. 211) partly made up of a dense yellowish or bluish opal that pervades brecciated and highly altered rocks with numerous open spaces of more than 10 cm in diameter. In some cases the alteration produced intensely yellow-brown or dark red tinted ocher as a powder.

Similar to the rocks at EP 91, the rock association of this hill is also interpreted as a fossil hydrothermal vent. However, because of the high content of opal at this location, the association is interpreted to have formed at a "white smoker" which is dominated by sulfate output.

EP 94

Pillow basalt, dikes, and ancient copper mine near Al Ghizayn
Topic: Formation of pillow basalts and dikes, copper smelting
Location: UTM 40 Q 498828 2634009 / N 23°49'02" E 56°59'18"

Rating: ☺☺

For location map see EP 93, Fig. 210.

Approach: Take the road #9 from Al Khabura (road #1) to Al Ghizayn. Watch out for the outcrop on the right hand side after about 21 km, just before entering the Wadi Hawasina and the village of Al Ghizayn.

The outcrop is along the steep cliff visible from the road to the north. Horizontally layered pillow basalts (formation of pillow basalts see Fig. 20 and EP 89), some of them showing chilled margins and altered rims of hyaloclastite, are intruded by vertical dikes (Figs. 212, 213). Massive basalt layers at the cliff top have columnar structures several meters long.

The flat area in front of the cliff is an archeological site where copper was mined around a thousand years ago. Unfortunately, the site has recently been severely damaged by construction work and may not be accessible in the future. Remnants of the

Fig. 212: Pillow basalt from a horizontal basalt layer at a steep cliff near Al Ghizayn.

Fig. 213: Dike intruded into horizontally layered pillow basalts near Al Ghizayn.

copper smelting can be detected on the ground in the form of little drop-shaped slag deposits. The metal comes from sulfide deposits in the vicinity of a black smoker in the neighborhood. Fossil black and white smokers can be observed at EP 91 and EP 93.

EP 95

Pillow basalts near Al Ghizayn
Topic: Formation of pillow basalts
Location: UTM 40 Q 498888 2633339 / N 23°48'40" E 56°59'20"

Rating: ☺

For location map see EP 93, Fig. 210.

Approach: Take the road #9 from Al Khabura (road #1) to Al Ghizayn. The outcrop appears directly beside the road on the left hand side, 500 m after the bridge into the Wadi Hawasina.

If there is no time to visit the world class outcrop of pillow basalts in Wadi al Jizzi (see EP 89) this outcrop may be an alternative to study some pillows of the upper part of the Samail Ophiolite. For explanation of the formation of pillow basalts see Fig. 20 and EP 89.

EP 96

Wadi ad Dil, contact Hawasina group – Samail Ophiolite
Topic: Sediments of the Hawasina group, metamorphic sole, overthrust, folds, schistosity
Locations: a) UTM 40 Q 491673 2617855 / N 23°40'34" E 56°55'10"
　　　　　b) UTM 40 Q 491673 2617855 / N 23°40'20" E 56°55'06"
　　　　　c) UTM 40 Q 491120 2617971 / N 23°40'22" E 56°54'46"
　　　　　d) UTM 40 Q 490757 2617389 / N 23°40'02" E 56°54'33"

Rating: ☺☺

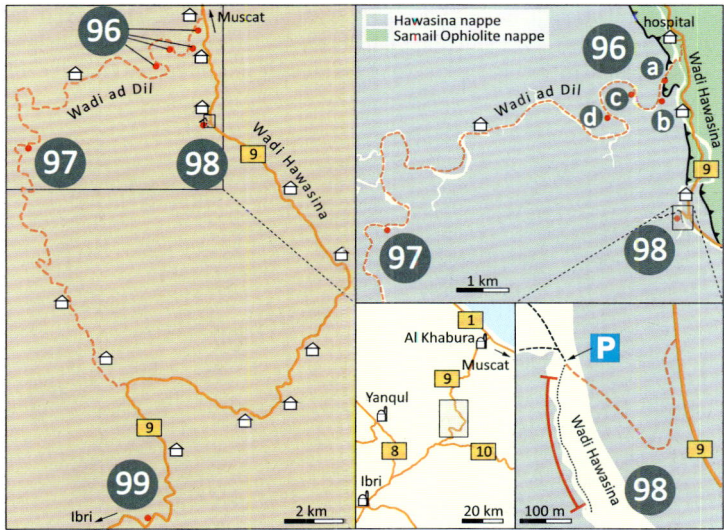

Fig. 214: Location map.

Approach: Take the road #9 from Al Khabura (road #1) to Al Ghizayn. Continue on road #9 for another 22 km. Pass the hospital on the right hand side and take the unpaved road about 100 m after the paved road to the parking area of the hospital. The unpaved road leads into Wadi ad Dil. Follow this road for 1 km to the first outcrop (EP 96a). Continue on the road for 600 m for the second (EP 96b) and another 800 m for the third outcrop (EP 96c).

The Wadi ad Dil, also named Wadi Dala, is a branch of the Wadi Hawasina and is mainly located within rocks of the Hawasina group. At the entrance of the wadi, the overthrust of the Samail Ophiolite onto the Hawasina group can be studied. In outcrop EP 96a (Fig. 214) highly deformed metamorphic rocks of the metamorphic sole are exposed (Fig. 215). For further explanation of the metamorphic sole see EP 37.

The Hawasina group formed at the northeastern edge of the Arabian Platform from Late Permian to Middle Cretaceous. It is composed of Permian to Triassic pelagic sediments and subordinate volcanic rocks in the upper Hawasina nappes and of Jurassic and Cretaceous base of slope and basin sediments in the lower Hawasina nappes which originally covered the upper tectonic units. The basin sediments of the lower unit are mainly characterised by deep marine sediments influenced by carbonate platforms in the surroundings. These are part of the Arabian Platform and include the Permian Baid and Triassic Misfah horst structures. Most of the sediments of the Hawasina basin are deposited below the carbonate compensation depth (CCD) and consist of turbidites, calcareous turbidites, radiolarites, and siliceous sediments. The depositional environments of the sediments are subdivided into 1. the distal Umar basin which is mainly composed of pelagic limestone and radiolarites since it was protected from the influx of

Fig. 215: Folded rocks of the metamorphic sole at the base of the Samail Ophiolite nappe in the Wadi ad Dil (EP 96a, Fig. 214).

Fig. 216: Folded radiolarian cherts of the Hawasina group in the Wadi ad Dil (EP 96b, Fig. 214).

Fig. 217: Folded turbiditic sequence of the Hawasina group in the Wadi ad Dil (EP 96c, Fig. 214).

detrital sediments by a topographic high (Kawr Ridge), and 2. the proximal Hamrat Duru basin, which is mainly filled with turbidites derived from the Arabian Platform and slope deposits.

In the Wadi ad Dil, Late Triassic to Jurassic radiolarites and turbiditic sequences of the Hamrat Duru basin as part of the Hawasina group, are exposed. Near the entrance into the wadi some outcrops of green and red radiolarian cherts can be studied (EP 96b, Fig. 214; Fig. 216), followed by turbiditic sequences (EP 96c, Fig. 214; Fig. 217) which

Fig. 218: Folded sediments of the Hawasina group in the Wadi ad Dil (EP 96d, Fig. 214).

Fig. 219: Folded turbidite sequence of the Hawasina group in the Wadi ad Dil with clearly visible bedding and cleavage.

are partly strongly folded (EP 96d, Fig. 214; Fig. 218). The sedimentary sequence is slightly metamorphosed and at some locations it clearly shows the difference between the inclination of bedding and cleavage (Fig. 219). Since in the case of Fig. 219, the bedding inclination is steeper than the cleavage, it belongs to the overturned limb of a larger fold structure which is not visible in its entirety (Fig. 220). However, a number of well exposed fold structures of dm to m scale can be observed at this location. Most of the fold structures are reclined folds (e.g., Fig. 218).

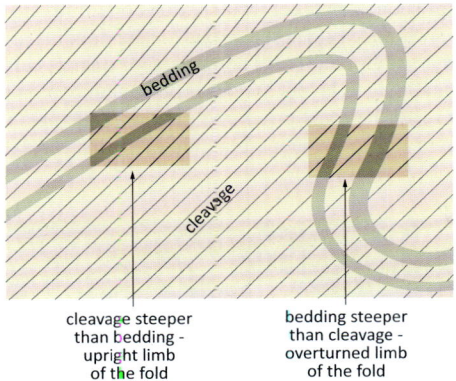

Fig. 220: Sketch of a vergent fold structure to explain the different angles between bedding and cleavage.

EP 97

Mullions and boudinage in sediments of the Hawasina group – the "mullion and boudinage museum"
Topic: Mullion structures and boudinage
Location: UTM 40 Q 486068 2614520 / N 23°38'28" E 56°51'48"

Rating: ☺☺☺

For location map see EP 96, Fig. 214.

Approach: Continue on the way into Wadi ad Dil, passing EP 97 and follow the unpaved road for another 6 km. The best outcrop is on the left hand/ southern side of the road; others are also visible on the right hand side.

At this outcrop a sequence of radiolaritic chert layers of the Hamrat Duru basin sediments is exposed. We call this outcrop a "mullion and boudinage museum" since a number of layers show the effects of boudinage and mullion structures within layers of different competencies in an extraordinary well exposed way (Fig. 221). Because of the contrast in competency of the various layers, mullion and boudinage structures developed which can be observed very clearly and also three-dimensionally at some places (Fig. 222).

Boudinage is a special expression of extensional deformation in layers of different competency. The term 'boudin' is a butcher's term, referring to long, parallel strands of sausages. In a vertical section parallel to the extensional direction, the boudins strikingly resemble the appearance of sausage strands. The boudins are separated by veins which are filled with crystal precipitates comprised of quartz, calcite, etc. If the originally extended and boudinaged layers are deformed again by compression in the same direction, the boudins may be transferred into mullion structures (Fig. 223). Thus, the mullion structures were formed as a combination of firstly, the extension with the for-

Fig. 221: Mullion structures in radiolaritic cherts of the Hamrat Duru basin.

Fig. 222: Detail of mullion structures, axis of mullions is vertical to the bedding.

Field sites 255

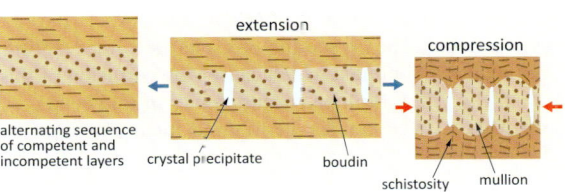

Fig. 223: Formation of boudins at extensive deformation and mullions at compressive deformation.

Fig. 224: Extensional boudinage, long axis of the boudins is parallel to the bedding.

mation of boudins (Fig. 224) and later, an overprinting by compression. The term 'mullion' comes from clustered columns in Gothic churches which support arches or divide lights of mullioned windows.

The long axes of the mullions are generally parallel to the fold plane of the fold in which they occur and thus vertical to the bedding in the fold hinge (Fig. 225). Mullions have a characteristic relation of length to height of about 1 : 2 which is in contrast to the

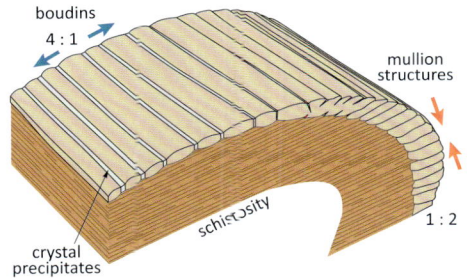

Fig. 225: Sketch of the development of mullion structures and boudins.

classical extensional boudinage with a relation of around 4 : 1 (Fig. 225), and a long axis parallel to the bedding. The crystalline precipitates formed during extension between the boudins act as an abutment which causes the extrusion of this rock material during compression (Fig. 223). Mullions occur mostly in fold bends, whereas the fold limbs may still display extensive boudinage. With an intense mullion formation, a secondary extensional boudinage may occur in the direction of the fold axis.

EP 98

Wadi Hawasina sediments
Topic: Sediments of the Hawasina group, colourful outcrop
Location: UTM 40 Q 492306 2615289 / N 23°38'53" E 56°55'28"

Rating: ☺

For location map see EP 96, Fig. 214.

Approach: Take the road #9 from Al Khabura (road #1) to Al Ghizayn. Continue on road #9 for another 26 km. After a small village on the right hand side, turn right (at UTM 40 Q 492456 2615476 / N 23°38'59" E 56°55'33") onto an unpaved road leading down into the Wadi Hawasina. The outcrop is well visible on the left hand side of the road.

Fig. 226: Sediments of the Hawasina group in the Wadi Hawasina.

Wadi Hawasina is orientated NNW-SSE here and follows the thrust contact between the serpentinised harzburgites of the Samail Ophiolite nappe to the east and the sediments of the Hawasina nappe to the west. Steeply inclined limestone and siliciclastic rocks of Triassic to Early Jurassic age that belong to the Matbat formation form the outcrop here (Villey et al. 1986a). The outcrop is particularly colourful, especially in the early morning hours (Fig. 226).

EP 99

Wadi al Kabir, Hawasina unit, coloured wall
Topic: Sediments of the Hawasina group
Location: UTM 40 Q 490333 2601318 / N 23°31'20" E 56°54'19"

Rating: ☺

For location map see EP 96, Fig. 214.

Approach: Take the road #10 from Ibri to Rustaq. Turn left into road #9 about 2 km after the village of Miskin (at UTM 40 Q 484613 2598563 / N 23°29'50" E 56°50'57"). The outcrop can be seen on both sides of the road after about 7 km.

Fig. 227: Sediments of the Hawasina group at a roadcut in the Wadi Al Kabir.

Another outcrop of rocks of the Hawasina group recently came into existence with the construction of the new road in the Wadi al Kabir in 2009. The sequence at the outcrop at this location belongs to the Hamrat Duru basin sediments of Jurassic age. The outcrop gives an insight into the layered sequences of the Hawasina group. The outcrop are along the road and show a multicoloured variety of cm to dm thick layers of sandstones, siltstones, shales and limestones interbedded with silty mudstones (Fig. 227). At some places deformations with folds and faults are well visible. For further information on the Hawasina group, see locations EP 96 to 98.

References

Al Azri, H. (1987): Typologie des gisements de chromite dans la partie sud de l'Ophiolite du Nord Oman. – Doctoral thesis, Orleans University. Bureau de Recherches Géologiques et Minières, Document n° 129, Orleans, France, 159 pp.

Al-Farraj, A. (2005): An evolutionary model for sabkha development on the north coast of the UAE. – Journal of Arid Environments, 63: 740.

Al Harthy, A. (2012): Palaeocene to Eocene mixed carbonate and siliciclastics sequences in Jabal Ja'alan. – Fieldtrip of the Geological Society of Oman, unpublished.

Al-Lazki, A., Seber, D., Sandvol, E., & Barazangi, M. (2002): A crustal transect across the Oman Mountains on the eastern margin of Arabia. – GeoArabia, 7(1): 47–78.

Allen, P.A. (2007): The Huqf Supergroup of Oman: basin development and context for Neoproterozoic glaciation. – Earth-Science Reviews, 84(3): 139–185.

Allen, P.A., Bowring, S., Leather, J., Brasier, M.D., Cozzi, A., Grotzinger, J.P., McCarron, G. & Amthor, J.E. (2002): Chronology of Neoproterozoic glaciations: New insights from Oman. – The 16th Int. Sediment. Congr., Abstr. Vol., Johannesburg, South Africa, 7–8.

Allen, P.A. & Etienne, J.L. (2008): Sedimentary challenge to Snowball Earth. – Nature Geoscience, 1: 817–825.

Anonymous (1972): Penrose Field Conference: Ophiolites. – Geotimes, 17: 24–25.

Armitage, S.J., Jasim, S.A., Marks, A.E., Parker, A.G., Usik, V.I. & Uerpmann, H.P. (2011): The southern route "out of Africa": evidence for an early expansion of modern humans into Arabia. – Science, 331: 453–456.

Arnaud, E., Halverson, G.P. & Shields-Zhou, G. (2011): Chapter 1: The geological record of Neoproterozoic ice ages. – Geol. Soc. London Mem., 36: 1–16.

Azzara, V.M. (2009): Domestic architecture at the Early Bronze Age sites HD-6 and RJ-2 (Jaalān, Sultanate of Oman). – Proceedings of the Seminar for Arabian Studies, 39: 1–16.

Batanouny, K.H. & Ismail, A.M. (1985) Phyto-ecological observations in northern Oman. – Qatar Univ. Sci. Bull., 5: 87–129.

Beavington-Penney, S.J., Wright, V.P. & Racey, A. (2006): The middle Eocene Seeb Formation of Oman: an investigation of acyclicity, stratigraphic completeness, and accumulation rates in shallow marine carbonate settings. – Journal of Sedimentary Research, 76(10): 1137–1161.

Bechennec, F., Le Metour, J., Rabu, D., Bourdillon-de-Grissac, C., de Wever, P., Beurrier, M. & Villey, M. (1990): The Hawasina Nappes: stratigraphy, palaeogeography and structural evolution of a fragment of the south-Tethyan passive continental margin. – In: Robertson, A., Searle, M. & Ries, A. (Ed.): The Geology and tectonics of the Oman region. – Geol. Soc. Spec. Publ., 49: 213–223.

Bendias, D., Schlaich, M. & Aigner, T. (2013): Reservoir and Seal Potential of the Mixed Carbonate-Siliciclastic Mafraq Fm., Oman: An Integrated Outcrop Analogue Study. – Extended abstract, Fifth EAGE Arabian Plate Geology Workshop 2015, Kuwait; DOI: 10.3997/2214-4609.201411948

Berger, J.F., Charpentier, V., Crassard, R., Martin, C., Davtian, G. & López-Sáez, J.A. (2013): The dynamics of mangrove ecosystems, changes in sea level and the strategies of Neolithic settlements along the coast of Oman (6000–3000 cal. BC)). – Journal of Archaeological Science, 40: 3087–3104.

Berner, R.A. (2006): GEOCARBSULF: A combined model for Phanerozoic atmospheric O_2 and CO_2. – Geochim. Cosmochim. Acta, 70: 5653–5664.

Beurrier, M., Bechennec, F., Rabu, D. & Hutin, G. (1986): Geological map of Rustaq, sheet NF 40-3D, 1:100,000, explanatory notes. – Sultanate of Oman, Ministry of Petroleum and Minerals, 69 pp.

Biagi, P. (1994a): A radiocarbon chronology for the aceramic shell-middens of coastal Oman. – Arabian Archaeology and Epigraphy, 5: 17–31.

Biagi, P. (1994b): An early Palaeolithic site near Saiwan (Sultanate of Oman). – Arabian archaeology and Epigraphy, 5: 81–88.

Blechschmidt, I., Matter, A., Preusser, F. & Rieke-Zapp, D. (2009): Monsoon triggered formation of Quaternary alluvial megafans in the interior of Oman. – Geomorphology, 110 (3–4): 128–139.

Blendinger, W., Van Vliet, A. & Clarke, M.H. (1990): Updoming, rifting and continental margin development during the Late Palaeozoic in northern Oman. – Geological Society, London, Special Publications, 49(1): 27–37.

Blount, C., Fritz, H.M. & Al-Harthy, A.H.M. (2010): Coastal vulnerability assessment based on historic tropical cyclones in the Arabian Sea. – In: Charabi, Y. (ed.): Indian Ocean Tropical Cyclones and Climate Change. Springer, New York, 295–303.

Boudier, F., Ceulener G. & Nicolas, A. (1988): Shear zones, thrusts and related magmatism in the Oman ophiolite: initiation of thrusting on an oceanic ridge. – Tectonophysics, 151: 275–296.

Boudier, F. & Nicolas, A. (1995): Nature of the Moho Transition Zone in the Oman Ophiolite. – Journal of Petrology, 36 (3): 777–796.

Buerkert, A. & Schlecht, E. (eds.) (2010): Oases of Oman: Millenia old livelihood systems at the crossroads. – Al Roya Press and Publishing House, Muscat, Oman, 136 pp.

Bourget, J., Zaragosi, S., Rodriguez, M., Fournier, M., Garlan, T. & Chamot-Rooke, N. (2013): Late Quaternary megaturbidites of the Indus Fan: Origin and stratigraphic significance. – Marine Geology, 336, 10–23.

Brongniard, A. (1813): Essai de classification des roches mélangées. – Journal des Mines, Paris, 199: 177–238.

Byrne, D.E. & Sykes, L.R. (1992): Great thrust earthquakes and aseismic slip along the plate boundary. – Journal of Geophysical Research, 97(B1): 449–478.

Cartwright, C.R. & Glover, E. (2002): Ra's al Hadd: Reconstructing the coastal environment in the 3rd millennium BC and later. – Journal of Oman Studies, 12: 41–53.

Central Bank of Oman (2015): Annual Report 2014. – available at http://www.cbo-oman.org/annual/CBOAnnualReport2014ENG.pdf

Charpentier, V. (2008): Hunter-gatherers of the "empty quarter of the early Holocene" to the last Neolithic societies: chronology of the late prehistory of south-eastern Arabia (8000–3100 BC). – Proceedings of the Seminar for Arabian Studies, 38(2008): 59–82.

Chavagnac, V., Ceuleneer, G., Monnin, C., Lansac, B., Hoareau, G. & Boulart C. (2013): Mineralogical assemblages forming at hyperalkaline warm springs hosted on ultramafic rocks: A case study of Oman and Ligurian ophiolites. – Geochemistry, Geophysics, Geosystems, 14: 2474–2495, doi:10.1002/ggge.20146

Clark, Sir T. (2007): Underground to Overseas: The Story of Petroleum Development Oman. – Stacey International, London, Great Britain. 123 pp.

Clarke, H. M.W. (1988): Stratigraphy and rock units nomenclature in the oil producing area of interior Oman. – Journal of Petroleum Geology, 11(1): 5–60.

Cleuziou, S. & Tosi, M. (2007a): The Search for Oman's Earliest Humans. – In: Cleuziou, S. & Tosi, M. (ed.): In the Shadow of the Ancestors. The Prehistoric Foundations of the Early Arabian Civilization in Oman. Ministry of the Heritage and Culture, Sultanate of Oman, 19–31. Al Nahda Printing Press.

Cleuziou, S. & Tosi, M. (2007b): The Great Transformation. – In: Cleuziou, S. & Tosi, M. (eds.): In the Shadow of the Ancestors. The Prehistoric Foundations of the Early Arabian Civilization in Oman. Ministry of the Heritage and Culture, Sultanate of Oman, 63–97. Al Nahda Printing Press.

Cleuziou, S. & Tosi, M. (2007c): A Greater Society Looms Under the Eyes of the Ancestors. – In: Cleuziou, S. & Tosi, M. (eds.): In the Shadow of the Ancestors. The Prehistoric Foundations of the Early Arabian Civilization in Oman. Ministry of the Heritage and Culture, Sultanate of Oman, 107–132. Al Nahda Printing Press.

Cleuziou, S. & Tosi, M. (2007d): Early Arabian Cilvilsation at its Zenith. – In: Cleuziou, S. & Tosi, M. (eds.): In the Shadow of the Ancestors. The Prehistoric Foundations of the Early Arabian Civilization in Oman. Ministry of the Heritage and Culture, Sultanate of Oman, 213–247. Al Nahda Printing Press.

Cleuziou, S. & Tosi, M. (2007e): Collapse and Transformation. – In: Cleuziou, S. & Tosi, M. (eds.): In the Shadow of the Ancestors. The Prehistoric Foundations of the Early Arabian Civilization in Oman. Ministry of the Heritage and Culture, Sultanate of Oman, 257–276. Al Nahda Printing Press.

Cleuziou, S. & Tosi, M. (2007f): The Iron Age: New Development on the Eve of History. – In: Cleuziou, S. & Tosi, M. (eds.): In the Shadow of the Ancestors. The Prehistoric Foundations of the Early Arabian Civilization in Oman. Ministry of the Heritage and Culture, Sultanate of Oman, 281–299. Al Nahda Printing Press.

Coleman, R.G. (1971): Plate tectonic emplacement of upper mantle peridotites along continental edges. – Journal of Geophysical Research, 76:1212–1222.

Copley, A., Avouac, J.-P. & Royer, J.-Y. (2010): India-Asia collision and the Cenozoic slowdown of the Indian plate: implications for the forces driving plate motions. – Journal of Geophysical Research, 115: B03410, doi:10.1029/2009JB006634.

Costa, P.M. & Wilkinson, T.J. (1987): The Hinterland of Sohar. – Journal of Oman Studies, 9: 9–238.

Cremarschi, M., Zerboni, A., Charpentier, V., Crassard, R., Isola, Il, Regattieri, E. & Zanchetta, G. (2015): Early-Middle Holocene environmental changes and pre-Neolithic human occupations as recorded in the cavities of Jebel Qara (Dhofar, southern Sultanate of Oman). – Quaternary International, 382(24): 264–276.

Dabirian, R., Beiranvand, M.S. & Aghahoseini, S. (2012): Mineral carbonation in peridotite rock for CO_2 sequestration and a method of leakage reduction of CO_2 in the rock. – Nafta, 63(1–2): 44–48.

Dale, A. & Hadwin, J. (2001): Adventure Trekking in Oman. – Hadwin D. and Yusuf, 34 Mile-End Avenue, Aberdeen, ISBN: 0953785408.

Davison, D. (1990): Meeting Place of the Spirits. – Aramco World, 1 (5, Sept–Oct).

De Gramont, X., Le Metour, J. and Villey, M. (1986): Explanatory notes. Geological Map of Samad. Sheet NF 40-7C. – Ministry of Petroleum and Minerals; Directorate General of Minerals.

Dennell, R. (2003): Dispersal and Colonizstion, long and short-term chronologies: how continuous is the Early Pleistocene record for hominids outside East Africa? – Journal of Human Evolution, 45: 421–440.

Dobretsov, S., Abed, R.M., Al Maskari, S.M., Al Sabahi, J.N. & Victor, R. (2011): Cyanobacterial mats from hot springs produce antimicrobial compounds and quorumsensing inhibitors under natural conditions. – Journal of Applied Phycology, 23(6): 983–993.

Drake, N.A., Blench, R.M., Armitage, S.J., Bristow, C.S. & White, K.H. (2011): Ancient watercourses and biogeography of the Sahara explain the peopling of the desert. – Proceedings of the National academy of Sciences, 108(2): 458–462.

Droste, H. & van Steenwinkel, M. (2004): Stratal geometries and patterns of platform Carbonates: The Cretaceous of Oman. – In: Seismic imaging of carbonate reservoirs and systems: AAPG Memoir, 81: 185–206.

El-Shazly, A.K., Bröcker, M., Hacker, B. & Calvert, A. (2001): Formation and exhumation of blueschists and eclogites from NE Oman: new perspectives from Rb–Sr and 40Ar/39Ar dating. – Journal of metamorphic Geology, 19: 233–248.

El-Shazly, A.K. & Coleman, R.G. (1990): Metamorphism in the Oman Mountains in relation to the ophiolite emplacement. – In Robertson, A.H.F., Searle, M.P. and Ries, A. (eds.): The Geology and Tectonics of the Oman Region: Geological Society London, Special Publication, 49: 475–495.

El-Shazly, A.K. & Liou, J.G. (1991): Glaucophane chloritoid bearing assemblages from NE Oman: Petrologic significance and a petrogenetic grid for high P/T metapelites. – Contributions Mineralogy and Petrology, 107: 180–201.

Enzel, Y., Amit, R., Dayan, U., Crouvi, O., Kahana, R., Ziv, B. & Sharon, D. (2008): The climatic and physiographic controls of the eastern Mediterranean over the late Pleistocene climates in the southern Levant and its neighboring deserts. – Global and Planetary Change, 60(3): 165–192.

Ernst, W.G. (1981): Petrotectonic setting of glaucophane schist belts and some implications for Taiwan. – Geological Society of China, Memoir, 4: 229–267.

Evans, A.M. (1993): Ore Geology and Industrial Minerals – an Introduction. – 3rd Ed., Blackwell Scientific Publications, Oxford, 400 pp.

Fairchild, I.J. & Kennedy, M. (2007): Neoproterozoic glaciation in the Earth System. – Journal of the Geological Society, London, 164: 895–921.

Filbrandt, J.B., Nolan, S.C. & Ries, A.C. (1990): Late Cretaceous and Early Tertiary evolution of Jebel Ja'alan and adjacent areas. – In: Robertson, A.H.F., Searle, M.P. & Ries, A. (eds.): The Geology and Tectonics of the Oman Region. Geological Society of London Special Publications, 49: 697–714.

Fleet, A.T. & Robertson, A.H.F. (1980): Ocean-ridge metalliferous and pelagic sediments of the Semail Nappe, Oman. – Journal of the Geological Society, 137(4): 403–422.

Fleitmann, D., Burns, S.J., Neff, U., Mangini, A. & Matter, A. (2003): Changing moisture sources over the last 330,000 years in Northern Oman from fluid-inclusion evidence in speleothems. – Quaternary Research, 60: 223–232.

Fleitmann, D. & Matter, A. (2009): The speleothem record of climate variability in Southern Arabia. – Comptes Rendus Geoscience, 341: 633–642.
Forbes G.A., Jansen H.S.M. & Schreurs J. (2010): Lexicon of Oman subsurface stratigraphy: reference guide to the stratigraphy of Oman's hydrocarbon basins. – GeoArabia Special Publication, 5: Gulf PetroLink, Bahrain.
Ford, T.D. & Pedley, H.M. (1996): A review of tufa and travertine deposits of the world. – Earth-Science Reviews, 41(3): 117–175.
Fournier, M., Chamot-Rooke, N., Petit, C., Fabbri, O., Huchon, P., Maillot, B. & Lepvrier, C. (2008): In situ evidence for dextral active motion at the Arabia–India plate boundary. – Nature Geoscience, 1(1): 54–58.
Fournier, M., Chamot-Rooke, N., Rodriguez, M., Huchon, P., Petit, C., Beslier, M.O. & Zaragosi, S. (2011): Owen fracture zone: the Arabia–India plate boundary unveiled. – Earth and Planetary Science Letters, 302(1): 247–252.
Fournier, M., Lepvrier, C., Razin, P. & Jolivet, L. (2006): Late Cretaceous to Paleogene post-obduction extension and subsequent Neogene compression in Oman Mountains. – GeoArabia, 11(4): 17–40.
Frisch, W., Meschede, M. & Blakey, R.C. (2011): Plate Tectonics. – Springer, Heidelberg Dordrecht London New York, 212 pp.
Frisch, W. & Meschede, M. (2013): Plattentektonik und Gebirgsbildung – 5th ed. (1st ed. 2005), Wissenschaftliche Buchgesellschaft/Primus-Verlag, Darmstadt. 196 pp.
Fritz, H.M., Blount, C.D., Albusaidi, F.B. & Al-Harthy, A.H.M. (2010): Cyclone Gonu storm surge in Oman. – Estuarine, Coastal and Shelf Science, 86(1): 102–106.
Gabunia, L., Vekua, A., Lordkipnidze, D., Swisher III, C.C., Ferring, R., Justus, A., Nioradze, M., Tvalchrelidze, M., Antón, S.C., Bosinski, G., Jöris, O., de Lumley, M-A., Majsuradze, G. & Mouskhelishvili, A. (2000): Earliest Pleistocene Hominid Cranial Remains from Dmanisi, Republic of Georgia: Taxonomy, Geological Setting and Age. – Science, 288 (5468): 1019–1025, DOI: 10.1126/science.288.5468.1019.
Gaidos, E.J., Nealson, K.H. & Kirschvink, J.L. (1999): Life in ice-covered oceans. – Science, 284(5,420): 1631–1633.
Gardner, R.A.M. (1988): Aeolianites and marine deposits of the Wahiba sands: character and palaeoenvironments. – Journal of Oman Studies Special Report, 3: 75–94.
Gass, I.G., Ries, A.C., Shackleton, R.M. & Smewing, J.D. (1990): Tectonics, geochronology and geochemistry of the Precambrian rocks of Oman. – Geological Society of London, Special Publications, 49(1): 585–599.
Ghazanfar, S.A. (1991): Vegetation structure and phytogeography of Jabal Shams, an arid mountain in Oman. – Journal of Biogeography, 18: 299–309.
Ghazanfar, S.A. (1999): A review of the flora of Oman. – In: Fisher, M., Ghazanfar, S.A. & Spalton, A. (eds.): The Natural History of Oman. A Festschrift for Michael Gallagher, pp. 29–63, Leiden: Backhuys Publ.
Glennie, K.W. (2005): The geology of the Oman Mountains – An outline of their history. – 110 pp., Beaconsfield (Scientific Press).
Glennie, K.W., Boeuf, M.G.A., Hugues-Clarke, M.W., Moody-Stuart, M., Pilaar, W.F.H. & Reinhardt, B.M. (1973): Late Cretaceous nappes in Oman Mountains and

their geologic evolution. – American Association of Petroleum Geologists, Bulletin, 57(1): 5–27.

Glennie, K.W., Boeuf, M.G.A., Hugues-Clark, M.H., Moody-Stuart, M., Pilaar, W.F.H. & Reinhardt, B.M. (1974): Geology of the Oman Mountains. – Mijnbouwkundig Genootschap, Deel 31: 423 pp.

Glennie, K.W., Clarke, M.H., Boeuf, M.G.A., Pilaar, W.F.H. & Reinhardt, B.M. (1990): Inter-relationship of Makran-Oman Mountains belts of convergence. – Geological Society of London, Special Publications, 49(1): 773–786.

Glennie, K.W. & Singhvi, A.K. (2002): Event stratigraphy, paleoenvironment and chronology of SE Arabian deserts. – Quaternary Science Reviews 21: 853–869.

Goldblatt, C., Lenton, T.M. & Watson, A.J. (2006): Bistability of atmospheric oxygen and the Great Oxidation. – Nature, 443: 683–686.

Goodenough, K.M., Thomas, R.J., Styles M.T, Schonfield, D.I. & MacLeod, C.J. (2014): Records of ocean growth and destruction in the Oman–UAE ophiolite. – Elements, 10: 109–114.

Grantham, P.J., Lijmbach, G.W.M., Posthuma, J., Hugues-Clarke, M.W. & Willink, R.J. (1988): Origin of crude oils in Oman. – Journal of Petroleum Geology, 11: 61–80.

Gray, D.R. & Gregory, R.T. (2003): Ophiolite obduction and the Samail Ophiolite: the behaviour of the underlying margin. – Geological Society of London, Special Publications, 218: 449–465.

Gray, D.R., Miller, J.M. & Gregory, R.T. (2005): Strain state and kinematic evolution of a fold-nappe beneath the Samail Ophiolite, Oman. – Journal of Structural Geology, 27: 1986–2007.

Grégoire, M., Langlade, J.A., Delpech, G., Dantas, C. & Ceuleneer, G. (2009): Nature and evolution of the lithospheric mantle beneath the passive margin of East Oman: evidence from mantle xenoliths sampled by Cenozoic alkaline lavas. – Lithos, 112(3): 203–216.

Grimes, C.B., Ushikubo, T., Kozdon, R. & Valley, J.W. (2013): Perspectives on the origin of plagiogranite in ophiolites from oxygen isotopes in zircon. – Lithos, 179: 48–66.

Grosjean, E., Love, G.D., Stalvies, C., Fike, D.A. & Summons, R.E. (2009): Origin of petroleum in the Neoproterozoic–Cambrian South Oman salt basin. – Organic Geochemistry, 40: 87–110.

Halverson, G.P., Hoffman, P.F., Schrag, D.P., Maloof, A.C. & Rice, A.H.N. (2005): Toward a Neoproterozoic composite carbon-isotope record. – Geological Society of America, Bulletin, 117: 1181–1207.

Hanna, S.S. (1995): Field guide to the Geology of Oman. – The Historical Association of Oman, International Printing Press, Ruwi, Oman, 178 pp.

Hannss, C. (1998): Geomorphologic characteristics of the configuration of the coastal landscape of Ra's al had in relation to some important relief features of Oman. – In: Hannss, C. & Kürschner, H. (ed.): The Capital Area of Northern Oman Teil II: Beihefte zum Tübinger Atlas des vorderen Orients, Reihe A, 31(2): 125–157.

Hannss, C. & Kürschner, H. (1998): The Capital Area of Northern Oman Teil II. – Beihefte zum Tübinger Atlas des vorderen Orients, Reihe A, 31(2): 125–157.

Harada, M. Tajika, E. & Sekine, Y. (2015): Transition to an oxygen-rich atmosphere with an extensive overshoot triggered by the Paleoproterozoic snowball Earth. – Earth and Planetary Science Letters, 419: 178–186.

Harland, W.B. (1964): Critical evidence for a great infra-Cambrian glaciation. – International Journal of Earth Sciences, 54: 45–61.

Hempton, M.R. (1987): Constraints on Arabian plate motion and extensional history of the Red Sea. – Tectonics, 6: 687–705. doi:10.1029/TC006i006p00687.

Heward, A. (2012): Geological Fieldtrip to Wadi Daiqa. – Field Guide of the Geological Society of Oman, unpublished.

Heward, A.P. & Penney, R.A. (2014): Al Khlata glacial deposits in the Oman Mountains and their implications. – Geological Society of London, Special Publications, 392: 279–301.

Hilbert, Y. H. & Azzara, V.M. (2011): Lithic technology and spatial distribution of artefacts at the Early Bronze Age site HD-6 (Sharqiyya Region, Sultanate of Oman). – Arabian archaeology and epigraphy, 22: 1–19.

Hoffmann, G. & Reicherter, K. (2014): Reconstructing Anthropocene extreme flood events by using litter deposits. – Global and Planetary Change, 122: 23–28.

Hoffmann, G., Grützner, C., Reicherter, K. & Preusser, F. (2015): Geo-archaeological evidence for a Holocene extreme flooding event within the Arabian Sea (Ras al Hadd, Oman). – Quaternary Science Reviews, 113: 123–133.

Hoffmann, G., Reicherter, K., Wiatr, T., Grützner, C. & Rausch, T. (2013a): Block and boulder accumulations along the coastline between Fins and Sur (Sultanate of Oman): tsunamigenic remains? – Natural Hazards, 65: 851–873. doi: 10.1007/s11069-012-0399-7.

Hoffmann, G., Rupprechter, M., Al Balushi, N., Grützner, C. and Reicherter, K. (2013b): The impact of the 1945 Makran tsunami along the coastlines of the Arabian Sea (Northern Indian Ocean) – a review – Zeitschrift für Geomorphologie, 57, Suppl. 4: 257–277. DOI: 10.1127/0372-8854/2013/S-00134.

Hoffmann, G., Rupprechter, M. & Mayerhofer, C. (2013c): Review of the longterm coastal evolution of Oman – subsidence versus uplift. – Zeitschrift der deutschen Gesellschaft für Geowissenschaften, 164: 237–252. DOI: 10.1127/1860-1804/2013/0002.

Hoffmann, G., Rupprechter, M., Rahn, M. & Preusser, F. (2015). Fluvio-lacustrine deposits reveal precipitation pattern in SE Arabia during early MIS 3. – Quaternary International, 382: 145–153.

Hoffman, P.F., Kaufman, A.J., Halverson, G.P. & Schrag, D.P. (1998): A Neoproterozoic snowball Earth. – Science, 281: 1342–1346.

Homewood, P., Razin, P., Grelaud, C., Droste, H., Vahrenkamp, V., Mettraux, M. & Mattner, J. (2008): Outcrop sedimentology of the Natih Formation, northern Oman: A field guide to selected outcrops in the Adam Foothills and Al Jabal al Akhdar areas. – GeoArabia, 13(3): 39–120.

Homewood, P., Vahrenkamp, V., Mettraux, M., Mattner, J., Vlaswinkel, B., Droste, H. & Kwarteng, A. (2007): Bar Al Hikman: a modern carbonate and outcrop analogue in Oman for Middle East Cretaceous fields. – First Break, 25: 55–61.

Hutin, G., Bechennec, F., Beurrier, M. & Rabu, D. (1986): Geological map of Birkat al Mawz: Sheet NF 40-7B, scale: 1:100,000, explanatory notes.

Immenhauser, A., Schreurs, G., Gnos, E., Oterdoom, H.W. & Hartmann, B. (2000): Late Palaeozoic to Neogene geodynamic evolution of the northeastern Oman margin. – Geological Magazine, 137(01): 1–18.

Immenhauser, A., Schreurs, G., Peters, T., Matter, A., Hauser, M. & Dumitrica, P. (1998): Stratigraphy, sedimentology and depositional environments of the Permian to uppermost Cretaceous Batain Group, eastern-Oman. – Eclogae geologicae Helvetiaem, 91: 217–235.

Jagher, R. (2009): The Central Oman Paleolithic Survey: Recent Research in Southern Arabia and Reflection on the Prehistoric Evidences. – In: Petraglia, M.D. & Rose, J.I. (eds.): The Evolution of Human Populations in Arabia: Paleoenvironments, Prehistory and Genetics – Vertebrate Paleobiology and Paleoanthropology Series, 139–150; Springer Netherlands.

Janjou, D., Minoux, L., Beurrier, M., De Gramont X., Le Métour; J. & Villey, M. (1986): Ibri – Sheet NF40-2F. Geological Map. Oman 1:100,000. – Ministry of Petroleum and Minerals, Directorate General of Minerals, Muscat.

Karpoff, A.M., Walter, A.V. & Pflumio, C. (1988): Metalliferous sediments within lava sequences of the Sumail ophiolite (Oman): Mineralogical and geochemical characterization, origin and evolution. – Tectonophysics, 151(1): 223–245.

Kelemen, P.B. & Matter, J. (2008): In situ carbonation of peridotite for CO_2 storage. – Proceeding of the National Academy of Science. U.S.A., 105(45) 17,295–17,300.

Kelemen, P.B., Matter, J., Streit, E.E., Rudge, J.F., Curry, W.B. & Blusztajn, J. (2011): Rates and mechanisms of mineral carbonation in peridotite: natural processes and recipes for enhanced, in situ CO_2 capture and storage. – Annual Review of Earth and Planetary Sciences, 39: 545–576.

Kellerhals, P. (1998): Ice age related erosional and depositional processes: examples from the late Proterozoic and Quaternary, Sultanate of Oman. – PhD Thesis, University Bern, 97 pp.

Kempf, O., Kellerhals, P., Lowrie, W. & Matter, A. (2000): Paleomagnetic directions in late Precambrian glaciomarine sediments of the Mirbat Sandstone Formation, Oman. – Earth and Planetary Science Letters, 175: 181–190.

Kennedy, M.J., Christie-Blick, N. & Sohl, L.E. (2001): Are Proterozoic cap carbonates and isotopic excursions a record of gas hydrate destabilization following Earth's coldest intervals? – Geology, 29: 443–446.

Kickmaier, W. & Peters, T. (1990): Manganese occurrences in the Al Hammah Range – Wahrah Formation, Oman Mountains. – Geological Society of London, Special Publications 49(1): 239–249.

Kilner, B., MacConall, N. & Brasier, M. (2005): Low-latitude glaciation in the Neoproterozoic of Oman. – Geology, 33: 413–416.

Kious, W. & Tilling, R.I. (1996): This Dynamic Earth. The Story of Plate Tectonics. – U.S. Government Printing Office, online edition: http://pubs.usgs.gov/gip/dynamic/

Kirschvink, J.L. (1992): Late Proterozoic low-latitude glaciation: the snowball Earth. – In: Schopf, J.W. & Klein, C. (eds.): The Proterozoic Biosphere, Cambridge University Press, Cambridge; 51–52.

Koehrer, B., Aigner, T., Forke, H. & Pöppelreiter, M. (2012): Middle to Upper Khuff (Sequences KS1 to KS4) outcrop-equivalents in the Oman Mountains: Grainstone architecture on a subregional scale. – GeoArabia, 17(4): 59–104.

Koepke, J., Schoenborn, S., Oelze, M., Wittmann, H., Feig, S.T., Hellebrand, E., Boudier, F. & Schoenberg, R. (2009): Petrogenesis of crustal wehrlites in the Oman ophiolite: Experiments and natural rocks. – Geochemistry, Geophysics, Geosystems, 10: Q10002, doi:10.1029/2009GC002488.

König, P. (2012): Plant life in the Umm as Samim, Oman – a case study in a major inland sabkha. – Journal of Arid Environments, 85: 122–127.

Kopp, C., Fruehn, J., Flueh, E.R.. Reichert, C., Kukowski, N., Bialas, J. & Klaeschen, D. (2000): Structure of the Makran subduction zone from wide-angle and reflection seismic data. – Tectonophysics, 329: 171–191.

Koster, B., Hoffmann, G., Grützner, C. & Reicherter, K. (2014): Ground penetrating radar facies of inferred tsunami deposits on the shores of the Arabian Sea (Northern Indian Ocean). – Marine Geology, 351: 13–24.

Knaust, D. (2009). Complex behavioural pattern as an aid to identify the producer of Zoophycos from the Middle Permian of Oman. – Lethaia, 42(2): 146–154.

Krumbiegel, G. & Krumbiegel, B. (1981): Fossilien der Erdgeschichte. – Enke, 406 pp.

Kukowski, N., Schillhorn, T., Flueh, E.R. & Huhn, K. (2000): Newly identified strike-slip plate boundary in the northeastern Arabian Sea. – Geology, 28: 355–358.

Kwarteng, A.Y., Dorvlo, A.S. & Vijaya Kumar, G.T. (2009): Analysis of a 27-year rainfall data (1977–2003) in the Sultanate of Oman. – International Journal of Climatology, 29: 605–617.

Lahr, M.M. & Foley, R. (1994) Multiple Dispersals and Modern Human Origins. – Evolutionary Anthropology: Issues, news, and reviews, 3(2): 48–60.

Larick, R. & Ciochon, R.L. (1996): The African emergence and Early Asian Dispersals of the Genus Homo. – American Scientist, 84: 538–551.

Leather, J., Allen, P.A., Brasier, M.D. & Cozzi, A. (2002): A Neoproterozoic snowball Earth under scrutiny: Evidence from the Fiq glaciation of Oman. – Geology, 30: 891–894.

Lees, G.M. (1928): The geology and tectonics of Oman and parts of SE Arabia. – Quarterly Journal of the Geological Society, 84(4): 585–670.

Le Guerroué, E., Allen, P.A. & Cozzi, A. (2005): Two distinct glacial successions in the Neoproterozoic of Oman. – GeoArabia, 10: 17–34.

Le Métour, J., Béchennec, F., Chèvremont, P., Roger, J. & Wyns, R. (1992): Geological Map of Oman, Sheet NG40-14 – Buraymi. – Sultanate of Oman, Ministry of Petroleum and Minerals.

Le Métour, J., Michel, J.C., Becnennec, F., Platel, J.P. & Roger, J. (1995): Geology and Mineral Wealth of the Sultanate of Oman. – Ministry of Petroleum and Minerals Directorate General of Minerals Sultanate of Oman, Muscat, 285pp.

Le Metour, J., Villey, M. & de Gramont, X. (1986a): Geological map of Masqat, sheet NF 40-4A, 1:100.000, explanatory notes. – Sultanate of Oman, Ministry of Petroleum and Minerals, 45 pp.

Le Metour, J., Villey, M. & de Gramont, X. (1986b): Geological map of Quryat, sheet NF 40-4D, 1:100.000, explanatory notes. – Sultanate of Oman, Ministry of Petroleum and Minerals, 72 pp.

Lijmbach, G.W.M., van der Veen, F.M. & Engelhardt, E.D. (1981): Geochemical characterisation of crude oils and source rocks using field ionisation mass spectrometry. – Advances in organic geochemistry, 788–798.

Lippard, S.J., Shelton, A.W. & Gass, Y.G. (1986): The ophiolite of northern Oman. – Geological Society of London, Memoir, 11: 178 pp.

Logers, L. (2015): Kupfer für die alte Welt. – Spektrum der Wissenschaften, April 2015: 66–70.

Loosveld, R.J., Bell, A. & Terken, J.J.M. (1996): The tectonic evolution of interior Oman. – GeoArabia, 1(1): 28–51.

Lordkipanidze, A., Vekua, A., Ferring, R., Rightmire, G.P., Zollikofer, C., Ponce de León, M.S., Augusti, J., Kiladze, G., Mouskhelishvili, A., Nioradze, M. & Tappen, M. (2006): A fourth hominin skull from Dmanisi, Georgia. – The Anatomical Record Part A, 288A(11): 1146–1157; doi: 10.1002/ar.a.20379.

Lordkipanidze, D., Jashashvili, T., Vekua, A., Ponce de León, M.S., Zollikofer, C.P.E., Rightmire, G.P., Pontzer, H., Ferring, R., Oms, O., Tappen, M., Bukhsianidze, M., Agusti, J., Kahlke, R., Kiladze, G., Martinez-Navarro, B., Mouskhelishvili, A., Nioradze, M. & Rook, L. (2007): Postcranial evidence from early Homo from Dmanisi, Georgia. – Nature, 449: 305–310; doi:10.1038/nature06134.

Lordkipanidze, D., Ponce de León, M.S., Margvelashvili, A., Rak, Y., Rightmire, G.P., Vekua, A. and Zollikofer, C.P.E. (2013): A Complete Skull from Dmanisi, Georgia, and the Evolutionary Biology of Early Homo. – Science, 342 (6156): 326–331; doi: 10.1126/science.1238484.

Lyakhovsky, V., Ben-Avraham, Z. & Achmon, M. (1994): The origin of the Dead Sea rift. – Tectonophysics, 240: 29–43.

MacLeod, C.J. & Yaouancq, G. (2000): A fossil melt lens in the Oman ophiolite: Implications for magma chamber processes at fast spreading ridges. – Earth and Planetary Science Letters, 176: 357–373.

Matter, J.M. & Kelemen, P.B. (2009): Permanent storage of carbon dioxide in geological reservoirs by mineral carbonation. – Nature Geosciences, 2(12): 837–841.

Mauz, B., Vacchi, M., Green, A., Hoffmann, G. & Cooper, C. (2015): Beachrock: a tool for reconstructing relative sea level in the far-field. – Marine Geology, 362: 1–16.

McClusky, S., Reilinger, R., Mahmoud, S., Sari, D.B. & Tealeb, A. (2003): GPS constraints on Africa (Nubia) and Arabia plate motions. – Geophysical Journal International, 155(1): 126–138.

McKay, C. (2000): Thickness of tropical ice and photosynthesis on a snowball Earth. – Geophysical Research Letters, 27(14): 2153–2156.

Méry, S. & Charpentier, V. (2013): Neolithic material cultures of Oman and the Gulf seashores from 5500–4500 BCE. – Arabian Archaeology and Epigraphy, 24: 73–78.

Miles, S.B. (1901): Across the Green Mountains of Oman. – The Geographical Journal, 18(5): 465–498.

Miller, A.G. & Morris, M. (1988): Plants of Dhofar. The southern region of Oman. Traditional, economic and medicinal uses. – 361 pp., Sultanate of Oman: Office for conservation of the environment.

Miller, J.M., Gray, D.R. & Gregory, R.T. (1998): Exhumation of high-pressure rocks in northeastern Oman. – Geology, 26: 235–238.

Miller, J.M., Gray, D.R. & Gregory, R.T. (2002): Geometry and significance of internal windows and regional isoclinal folds in northeast Saih Hatat, Sultanate of Oman. – Journal of Structural Geology, 24: 359–386.

Millson, J.A., Mercadier, C.G.L., Livera, S.E. & Peters, J.M. (1996): The Lower Palaeozoic of Oman and its context in the evolution of a Gondwanan continental margin. – Journal of the Geological Society, 153(2): 213–230.

Ministry of Regional Municipalities, Environment & Water Resources (2006): The Aflaj irrigation system of Oman. Nomination of the UNESCO World Heritage. 33pp. – Retrieved from: http://whc.unesco.org/uploads/nominations/1207.pdf

Morton, D.M. (2006): In the Heart of the Desert: The Story of an Exploration Geologist and the Search for Oil in the Middle East. – Green Mountain Press, Aylesford, England.

Murty, T.S. & El-Sabh, M.I. (1984): Cyclones and strom surges in the Arabian Sea: A brief review. – Deep Sea Research Part A. Oceanographic Research Papers, 31(6): 665–670.

Musson, R.M.W. (2009): Subduction in the Western Makran: the historian's contribution. – Journal of the Geological Society, 166: 387–391.

Nasir, S., Al-Sayigh, A., Alharthy, A. & Al-Lazki, A. (2006): Geochemistry and petrology of Tertiary volcanic rocks and related ultramafic xenoliths from the central and eastern Oman Mountains. – Lithos, 90: 249–270.

Nasir, S., Al-Khirbash, S., Rollinson, H., Al-Harthy, A., Al-Sayigh, A., Al-Lazki, A. & Al-Busaidi, S. (2008): Evolved carbonatitic kimberlites from the Batain Nappes, eastern Oman continental margin. – In: 9th Kimberlite Conference Extended abstract (No. 91KC-A, p. 00002).

Nasir, S., Al-Khirbash, S., Rollinson, H., Al-Harthy, A., Al-Sayigh, A., Al-Lazki, A., Theye, T., Massonne, H.-J. & Belousova, F. (2011): Petrogenesis of ultramafic lamprophyres and carbonatites from the Batain Nappes, eastern Oman continental margin. – Contributions to Mineralogy and Petrology, 161: 47–74.

National Centre of Statistics and Information (2014): Statistical Yearbook. – Sultanate of Oman, 42: 556 pp.

Nicolas, A. (1989): Structures of ophiolites and dynamics of oceanic lithosphere. – Kluwer Academic Publishers, 367 pp.

Nicolas, A. (1995): The Mid-Oceanic Ridges – Mountains Below Sea Level. – Springer, Berlin-Heidelberg, 217 pp.

Nicolas, A. (2016): Ophiolite. – In: Harff, J., Meschede, M., Petersen, S. & Thiede, J. (eds.): Encyclopedia of Marine Geosciences, Springer Science, Dordrecht; DOI 10.1007/978-94-007-6644-0_127-1.

Nicolas, A. & Al Azri, H., 1991, Chromite-Rich and Chromite-Poor Ophiolite: The Oman Case. – In: Peters, T., Nicolas, A & Coleman, R.G. (eds.): Ophiolite Genesis and Evolution of the Oceanic Lithosphere Proceedings of the Ophiolite Conference (Muscat, Oman, 1990): Kluwer Academic Publishers, Dordrecht/Boston/London, 261–274.

Page, W.D., Alt, J.N., Cluff, L.S & Plafker, G. (1979): Evidence for the recurrence of large-magnitude earthquakes along the Makran coast of Iran and Pakistan. – Tectonophysics, 52(1): 533–547.

Parker, A.G. (2009): Pleistocene climate change in Arabia: developing a framework for Hominin dispersal over the last 350 ka. – In: Petraglia, M.D. & Rose, J.I. (eds.): The Evolution of Human Populations in Arabia. Springer, London, 39–51.

Parker, A.G. & Rose, J.I. (2008): Climate change and human origins in southern Arabia. – Proceedings of the Seminar for Arabian Studies, 38: 25–42.

Parton, A., White, T.S., Parker, A.G., Breeze, P.S., Jennings, R., Groucutt, H.S. & Petraglia, M.D. (2015): Orbital-scale climate variability in Arabia as a potential motor for human dispersals. – Quaternary International, 382: 82–97.

Patzelt, A. (2011): The *Themeda quadrivalvis* tall-grass savannah of Oman at the crossroad between Africa and Asia. – Edinburgh Journal of Botany, 68: 301–319.

Peters, T., Al Battashy, M., Bläsi, H., Hauser, M., Immenhauser, A., Moser, L. & Al Rajhi, A. (2001): Geological map of Sur and Al Ashkharah, Sheet NF 40-8F and nF 40-12C, Explanatory Notes. – Sultanate of Oman, Ministry of Commerce and Industry: 95pp.

Peters, J.M., Filbrandt, J., Grotzinger, J., Newall, M., Shuster, M. & Al-Siyabi, H. (2003): Surface-piercing Salt Domes of interior North Oman, and their significance of the Ara Carbonate Stinger" hydrocarbon play. – GeoArabia Manama, 8: 231–270.

Petraglia, M.D. (2003): The Lower Paleolithic of the Arabian peninsula: occupations, adaptions, and dispersals. – Journal of World Prehistory, 17(2): 141–179.

Petraglia, M.D. (2011): Trailblazers across Arabia. – Nature, 470: 50–51.

Petraglia, M.D., Drake, N. and Alsharekh, A. (2009): Acheulean Landscapes and Large Cutting tools Assemblages in the Arabian peninsula. – In: Rose, J.I. & Petraglia, M.D. (eds.): The Evolution of Human Populations in Arabia: Paleoenvironments, Prehistory and Genetics – Vertebrate Paleobiology and Paleoanthropology Series, 103–116; Springer Netherlands.

Petraglia, M.D., Parton, A., Groucutt, H.S. & Alsharekh, A. (2015): Green Arabia: Human prehistory at the Crossroads of Continents. – Quaternary International, 382: 1–7.

Pickering, H. & Patzelt, A. (2008): Field guide to the wild plants of Oman. – 282 pp., Kew: Royal Botanic Gardens.

Pierrehumbert, R.T. (2002): The hydrologic cycle in deep-time climate problems. – Nature 419: 191–198.

Pierrehumbert, R.T., Abbot, D.S., Voigt, A. & Koll, D. (2011): Climate of the Neoproterozoic. – Annual Review of Earth Sciences, 39: 417–460.

Pillevuit, A., Marcoux, J., Stampfli, G. & Baud, A. (1997): The Oman Exotics: a key to the understanding of the Neotethyan geodynamic evolution. – Geodinamica Acta, 10(5): 209–238.

Preusser, F., Radies, D. & Matter, A. (2002): A 160,000-year record of dune development and atmospheric circulation in Southern Arabia. – Science, 296: 2018–2020.

Preusser, F., Radies, D., Driehorst, F. & Matter, A. (2005): Late Quaternary history of the coastal Wahiba sands, Sultanate of Oman. – Journal of Quaternary Science, 20: 395–405.

Prins, M.A. & Postma, G. (2000): Effects of climate, sea level, and tectonics unraveled for last deglaciation turbidite records of the Arabian Sea. – Geology, 28(4): 375–378.

Qidawi, H.A. (2004): Industrial Rocks and Minerals in the Sultanate of Oman, Development Possibilities, Directorate General of Minerals, Ministry of Commerse and Industry, 164 pp.

Rabu, D., Bechennec, F., Beurrier, M. & Hutin, G. (1986): Geological map of Nakhl, sheet NF 40-3e, scale: 1:100,000, explanatory notes. – Sultanate of Oman, Ministry of Petroleum and Minerals, Directorate General of Minerals.

Radies, D., Preusser, F., Matter, A. & Mange, M. (2004): Eustatic and climatic controls on the development of the Wahiba Sand Sea, Sultanate of Oman. – Sedimentology, 51: 1359–1385.

Radies, D., Hasiotis, S.T., Preusser, F., Neubert, E. & Matter, A. (2005): Paleoclimatic significance of Early Holocene faunal assemblages in wet interdune deposits of the Wahiba Sand Sea, Sultanate of Oman. – Journal of Arid Environments, 62: 109–125.

Rajendran, S., Al-Khirbash, S., Pracejus, B., Nasir, S., Al-Abri, A.H., Kusky, T.M. & Ghulam, A. (2012): ASTER detection of chromite bearing mineralized zones in Semail Ophiolite Massifs of the northern Oman Mountains: Exploration strategy. – Ore Geology Reviews, 44: 121–135.

Reade, J. (2000): Sacred places in ancient Oman. – Journal of Oman studies, 11: 133–138.

Regard, V., Bellier, O., Thomas, J.C., Bourle, D., Bonnet, S., Abbassi, M.R., Braucher, R., Mercier, J., Shabanian, E., Soleymani, S. & Feghhi, K. (2005): Cumulative right-lateral fault slip rate across the Zagros–Makran transfer zone: role of the Minab–Zendan fault system in accommodating Arabia–Eurasia convergence in southeast Iran. – Geophysical Journal International, 162(1): 177–203. doi:10.1111/j.1365-246X.2005.02558.x.

Reuning, L., Schoenherr, J., Heimann, A., Urai, J.L., Littke, R., Kukla, P.A. & Rawahi, Z. (2009): Constraints on the diagenesis, stratigraphy and internal dynamics of the surface-piercing salt domes in the Ghaba salt basin (Oman): a comparison to the Ara formation in the South Oman salt basin. – GeoArabia, 14(3): 83–120.

Rioux, M., Bowring, S., Kelemen, P., Gordon, S., Miller, R. & Dudas, F. (2013): Tectonic development of the Samail ophiolite: High precision U-Pb zircon geochronology and Sm-Nd isotopic constraints on crustal growth and emplacement. – Journal of Geophysical Research, 118: 2085–2101; doi:10.1002/jgrb50139.

Robertson, A.H.F. & Fleet, A.J. (1986): Geochemistry and palaeo-oceanography of metalliferous and pelagic sediments from the Late Cretaceous Oman ophiolite. – Marine and Petroleum Geology, 3(4): 315–337.

Robertson, A.F.H., Searle M.P. & Ries A.C. (eds.) (1990): The Geology and Tectonics of the Oman Region. – Geological Society, London, Special Publications, 49.

Rodriguez, M., Fournier, M., Chamot-Rooke, N., Huchon, P., Bourget, J., Sorbier, M. & Rabaute, A. (2011): Neotectonics of the Owen Fracture Zone (NW Indian Ocean): Structural evolution of an oceanic strike-slip plate boundary. – Geochemistry, Geophysics, Geosystems, 12 (12).

Roger, J., Bechennec, F., Janjou, D., Le Metour, J., Wyns, R. & Beurrier, M. (1991): Geological Map of Ja'alan, Sheet NF 40-8E, Explanatory notes. – Sultanate of Oman, Ministry of Commerce and Industry.

Rohling, E.J., Grant, K.M., Roberts, A.P. & Larrasoaña, J.C. (2013): Paleoclimate Variability in the Mediterranean and Red Sea Regions during the Last 500,000 Years. – Current Anthropology, 54(S8): S183–S201.

Rollinson, H. (2005): Chromite in the mantle section of the Oman ophiolite. – The Island Arc, 14: 542–550.

Rollinson, H.R., Searle, M.P., Abbasi, I.A., Al-Lazki, A.I. & Al Kindi, M.H. (2014): Tectonic evolution of the Oman Mountains. – Geological Society, London, Special Publication, 392.

Rose, J.I. & Petraglia, M.D. (2009): Tracking the Origin and Evolution of Human Populations in Arabia. – In: Rose, J.I. & Petraglia, M.D. (eds.): The Evolution of Human Populations in Arabia. – Vertebrate Paleobiology and Paleoanthropology Series, 1–12; Springer Netherlands.

Rose, J.I. & Usik, V. (2009): The "Upper Paleolithic" of South Arabia. – In: Petraglia, M.D. & Rose, J.I. (eds.): Evolution of Human Populations in Arabia: Paleoenvironments, Prehistory and Genetics, 169–185. Springer Netherlands.

Rose, J.I, Usik, V.I., Marks, A.E, Hilbert, Y.H., Galletti, C.S., Parton, A., Geiling, J.M., Černý, V., Morley, M.W. & Roberts, R.G. (2011): The Nubian Complex of Dhofar, Oman: An African Middle Stone Age Industry in Southern Arabia – PLoS ONE 6(11)e28239. doi:10.1371/journal.pone.0028239.

Rosenberg, T.M., Preusser, F., Fleitmann, D., Schwalb, A., Penkman, K., Schmid, T.W., Al-Shanti, M.A., Kadi, K. & Matter, A. (2011): Humid periods in southern Arabia: windows of opportunity for modern human dispersal. – Geology, 39: 1115–1118.

Rosenberg, T.M., Preusser, F., Blechschmidt, I., Fleitmann, D., Jagher, R. & Matter, A. (2012): Late Pleistocene palaeolake in the interior of Oman: a potential key area for the dispersal of anatomically modern humans out-of-Africa? – Journal of Quaternary Science, 27: 13–16.

Ross D.A., Uchupi E. & White R.S. (1986): The geology of the Persian Gulf–Gulf of Oman region: A synthesis. – Reviews of Geophysics, 24: 537–556. doi: 10.1029/RG024i003p00537.

Salvatori, S. (2007): The Prehistoric graveyard of Ra's al-Hamra RH-5. – In: Cleuziou, S. & Tosi, M. (eds.): In the Shadow of the Ancestors. The Prehistoric Foundations of the Early Arabian Civilization in Oman. Ministry of the Heritage and Culture, Sultanate of Oman, 98–102. Al Nahda Printing Press.

Schoenherr, J., Schléder, Z., Urai, J.L., Littke, R. & Kukla, P.A. (2010): Deformation mechanisms of deeply buried and surface-piercing Late Pre-Cambrian to Early Cambrian Ara Salt from interior Oman. – International Journal of Earth Sciences, 99(5): 1007–1025.

Schreuers, G. & Immenhauser, A. (1999): West-northwest directed obduction of the Batain Group on the eastern Oman continental margin at Cretaceous-Tertiary boundary. – Tectonics, 18: 148–160.

Schulp, A.S., Hanna, S.S., Hartman, A.F. & Jagt, J.W. (2000): A Late Cretaceous theropod caudal vertebra from the Sultanate of Oman. – Cretaceous Research, 21(6): 851–856.

Schulp, A.S., O'Connor, P.M., Weishampel, D.B., Al-Sayigh, A.R. & Al-Harthy, A. (2008): Ornithopod and sauropod dinosaur remains from the Maastrichtian A-Khod Conglomerate. – Sultanate of Oman. Sultan Qaboos University Journal of Science, 13: 27–32.

Searle, M.P. (2007): Structural geometry, style and timing of deformation in the Hawasina Window, Al Jabal al Akhdar and Saih Hatat culminations, Oman Mountain. – Geo-Arabia 12 (2): 99–130.

Searle, M.P. (2014): Preserving Oman's geological heritage: Proposal for establishment of world heritage sites, national geoparks and sites of special scientific interest (SSSI). – Geological Society of London, Special Publications, 392: 9–44.

Searle, M.P. & Alsop, G.I. (2007): Eye-to-eye with a mega-sheath fold: A case study from Wadi Mayh, northern Oman Mountains. – Geology, 35: 1043–1046.

Searle, M.P. & Cox, J. (1999): Tectonic setting, origin, and obduction of the Oman ophiolite. – Geological Society of America Bull., 111 (1): 104–122.

Searle, M.P. & Graham, G.M. (1982): "Oman Exotics" – Oceanic carbonate build-ups associated with the early stages of continental rifting. – Geology, 10: 43–49.

Searle, M.P., Waters, D.J., Martin, H.N. & Rex, D.C. (1994): Structure and metamorphism of blueschist – eclogite facies rocks from the northeastern Oman Mountains. – Journal of the Geological Society, London, 151: 555–576.

Smith, G., McNeill, L., Henstock, T.J. and Bull, J. (2012): The structure and fault activity of the Makran accretionary prism. – Journal of Geophysical Research: Solid Earth (1978–2012), 117(B7).

Smith, G.L., McNeill, L.C., Wang, K., He, J. & Henstock, T.J. (2013): Thermal structure and megathrust seismogenic potential of the Makran subduction zone. – Geophysical Research Letters, 40(8): 1528–1533.

Stringer, C. (2000): Coasting out of Africa. – Nature, 405: 24–26.

Taboroši, D. & Kázmér, M. (2013): Erosional and depositional textures and structures in coastal karst landscapes. – In: Lace, M.J. & Mylroie, J. (eds.): Coastal karst landforms, Coastal research library, 5: 15–57, Springer, Berlin.

Terken, J.M.J., Frewin, N.L. & Indrelid, S.L. (2001): Petroleum systems of Oman: charge timing and risks. – American Association of Petroleum Geologists, Bulletin, 85(10): 1817–1845.

Tosi, M. & Usai, D. (2003): Preliminary report on the excavations at Wadi Shab, Area 1, Sultanate of Oman. – Arabian Archaeology and Epigraphy, 14: 8–23.

Uchupi, E., Swift, S.A. & Ross, D.A. (2002): Tectonic geomorphology of the Gulf of Oman Basin. – Geological Society of London, Special Publication, 195: 37–70.

UNESCO (2014a): Archaeological Sites of Bat, Al-Khutm and Al-Ayn. – Retrieved from: http://whc.unesco.org/en/list/434.

UNESCO (2014b): Aflaj Irrigation System of Oman. – Retrieved from: http://whc.unesco.org/en/list/1207.

Usik, V.I., Rose, J.I., Hilber, Y.H., Van Peer P. & Marks, A.E. (2012): Nubian Complex reduction strategies in Dhofar, southern Oman. – In: Groucutt, H. & Blinkhorn, J. (eds.): The Middle Palaeolithic in the Desert. – Quaternary International, 300: 244–266.

Vekua, A., Lordkipanidze, D., Rightmire, G.P., Agusti, J., Ferring, R., Maisuradze, G., Mouskhelishvil, A., Nioradze, M., Ponce de Léon, M., Tappen, M., Tvalchrelidze, M. & Zollikofer, C. (2002): A New Skull of Early Homo from Dmanisi, Georgia. – Science, 297(5578): 85–89, DOI: 10.1126/science.1072953.

Van Peer, P. (1998): The Nile Corridor and the Out-of-Africa Model. An examination of the Archaeological Record. – Current Anthropology, 39: 115–140.

Vermeersch, P.M. (2001): 'Out of Africa' from an Egyptian point of view. – Quaternary International, 75: 103–112.

Vernant, P., Nilforoushan, F., Hatzfeld, D., Abbassi, M.R., Vigny, C. & Masson, F. (2004): Present-day crustal deformation and plate kinematics in the Middle East constrained by GPS measurements in Iran and northern Oman. – Geophysical Journal International, 157(1): 381–398. doi:10.1111/j.1365-246X.2004.02222.x.

Villey, M., Bechennec, F., Beurrier, M., Le Metour, J. & Rabu, D. (1986): Geological map of Yanqul, explanatory notes. – Directorate General of Minerals, Oman Ministry of Petroleum and Minerals, sheet NF 40-2C, scale 1:100000.

Villey, M., Le Metour, J. & de Gramont, X. (1986): Geological map of Fanja, sheet NF 40-3F, 1 : 100.000, explanatory notes. – Sultanate of Oman, Ministry of Petroleum and Minerals, 68 pp.

Vinx, R. (2005): Gesteinsbestimmung im Gelände. – 1st Ed., Elsevier Spektrum Akademischer Verlag, München, 442 pp.

Warburton, J., Burnhill, T.J., Graham, R.H. & Isaac, K.P. (1990): The evolution of the Oman Mountains foreland basin. – In: Robertson, A.H.F., Searle, M.P. & Ries, A.C. (eds.): The geology and tectonics of the Oman region, Geological Society, Special Publication, 49: 419–427.

Warren, C.J., Parrish, R.R., Waters, D.J. & Searle, M.P. (2005): Dating the geological history of Oman's Semail ophiolite: insights from U-Pb geochronology. – Contributions Mineralogy and Petrology, 150: 403–422.

Weidlich, O. & Bernecker, M. (2007). Differential severity of Permian–Triassic environmental changes on Tethyan shallow-water carbonate platforms. – Global and Planetary Change, 55(1): 209–235.

Weisgerber, G. (1987): Archaeological evidence of copper exploitation at Arja. – The Journal of Oman Studies, 9: 145–172.

Weisgerber, G. (1991): Die Suche nach dem altsumerischen Kupferland Makan. – Das Altertum 37: 76–90.

Weisgerber, G. (2007a): Copper from Magan for Mesopotamian Cities. – In: Cleuziou S. & Tosi, M. (eds.): In the Shadow of the Ancestors. The Prehistoric Foundations of the Early Arabian Civilization in Oman. Ministry of the Heritage and Culture, Sultanate of Oman, 195–196. Al Nahda Printing Press.

Weisgerber, G. (2007b): From green to red: smelting Red copper from green ore. – In Cleuziou, S. & Tosi, M. (eds.): In the Shadow of the Ancestors. The Prehistoric Foundations of the Early Arabian Civilization in Oman. Ministry of the Heritage and Culture, Sultanate of Oman, 197–198. Al Nahda Printing Press.

Weisgerber, G. (2007c): Iron age mining and smelting (Lizq Period. – In: Cleuziou, S. & Tosi, M. (eds.): In the Shadow of the Ancestors. The Prehistoric Foundations of the Early Arabian Civilization in Oman. Ministry of the Heritage and Culture, Sultanate of Oman, 302–303. Al Nahda Printing Press.

Weyhenmeyer, C.E., Burns, S.J., Waber, H.N., Aeschbach-Hertig, W., Kipfer, R., Loosli, H.H. & Matter, A. (2000): Cool glacial temperatures and changes in moisture source recorded in Oman groundwaters. – Science, 287: 842–845.

Whalen, N.M., Zoboroski M. & Schubert, K. (2002): The Lower Palaeolithic in Southwestern Oman. – Adumatu, 5: 27–34.

White, R.S. (1977): Recent fold development in the Gulf of Oman. – Earth and Planetary Science Letters, 36(1): 85–91.

Wilson, R.A. (1997): Geochemistry of metalliferous sediments from the northern Oman ophiolite. – Durham theses, Durham University. Available at Durham E-Theses Online: http://etheses.dur.ac.uk/4979/

Woods, W.W. & Imes, J.L. (1995): How wet is wet? Precipitation constraints on late Quaternary climate in the southern Arabian Peninsula. – Journal of Hydrology, 164: 263–268.

Wyns, R., Le Métour, J., Roger, J. & Chevrel, S. (1992): Geological map of Sur 1:250.000, Sheet NF 40-08, explanatory notes. – Ministry of Petroleum and Minerals, Directorate General of Minerals, Muscat, Oman, 80 pp.

Yule, P (2001): The Hasat Bani Salt in the al-Zahirah Province of the Sultanate of Oman. – 9 pp., Retrieved from: http://archiv.ub.uni-heidelberg.de/propylaeumdok/132.

Yule, P. (2008): Sasanian Presence and Late Iron Age Samad in Central Oman, some Corrections. – Schriften von Paul Yule zu Arabien, 14: http://archiv.ub.uni-heidelberg.de/propylaeumdok/volltexte/2008/121

Yule, P. & Bergoffen, C. (1999): East of Ibrī: A Jahil in the Sharqīyah. – Orient-Archäologie, 2: 187–193.

Yule, P. & Kervran, M. (1993): More than Samad in Oman: Iron Age pottery from Şuhār and Khor Rorī. – Arabian archeology and epigraphy, 4: 69–106.

Yule, P. & Weisgerber, G. (1998): Prehistoric tower tombs at Shir/Jaylah, Sultanate of Oman. – Beiträge zur allgemeinen und vergleichenden Archäologie, 18: 183–241.

Zaim, Y., Ciochon, R.L., Polanski, J.M., Grine, F.E., Bettis III, E.A., Rizal Y., Franciscus, R.G., Larick, R.R., Heizler, M., Aswan, Eaves, K.L. & Marsh, H.E. (2011): New 1.5 million-year-old Homo erectus maxilla from Sangiran (Central Java, Indonesia). – Journal of Human Evolutions, 61: 363–376.

Zhang, D.D., Zhang, Y., Zhu, A. & Cheng, X. (2001): Physical mechanisms of river waterfall tufa (travertine) formation. – Journal of Sedimentary Research, 71(1): 205–216.

Index list

Abu Mahara formation 36, 62, 173
Abu Tan sand 60
Acheulean 6, 7
Aeolian deposit 17, 132, 133, 151, 157
Aeolianite 131, 132, 133, 150, 151
African Plate 40, 41, 42, 187, 223
Ain Al Hamman 31
Al Aqur 203
Al Awabi 183, 184
Al Ayn 11, 12, 203, 217, 219, 220, 221
Al Ashkharah 149, 151
Al Buraymi 59
Al Dhakhiliyah 31
Al Ghizayn 245, 247, 248
Al Ghubaira 86
Al Haddah 124
Al Hajar 80, 83, 85
Al Hajar Mountains 18, 23, 25, 30, 31,
    37, 44, 45, 64, 130, 160, 190, 192,
    193, 194, 204, 223
Al Hamra 194, 195
Al Hatab 9
Al Hayl 217
Al Jebel al Akhdar 199, 202, 203, 204
Al Kamil 29, 127
Al Khoud 39
Al Kasfah 31
Al Khutm 12, 220
Al Muaydin 204
Al Ruwais 123
Al Wasit 233
Al Wusta 59
Albedo 52, 54, 188
Alluvial fan 40
Alpine Orogeny 37
Amdeh formation 36, 60, 83, 85, 87
Amethyst 115
Amphibolite 46, 51, 143
Anhydrite 161
Ar Rawdah 194
Ar Rustaq 31
Arabian Platform 36, 37, 188

Arabian Plate 40, 42, 44, 102, 130, 187,
    200, 204, 208
Arabian Sea 18
Arja 13, 237
As Sifah 44, 76, 77, 78, 81
As Siwayh 125, 127
Asthenosphere 47, 48
Asylah 126, 127
Ayn A' Thowarah 31, 164

Bab al Mendab strait 9
Baida 237
Bandar al Khayran 73, 75, 76
Bani Jabir plateau see Selma plateau
Bar Al Hikman 32
Barchan 147
Basalt 47, 49, 50, 51, 52, 226, 228
Basanite 149
Bat 12, 13, 220
Batain basin 44
Batain nappe 116, 118, 123, 129
Batinah 23, 28, 30, 40, 163, 194
Beach rock 99, 101
Beehive tomb 11, 97, 163, 218, 220,
    221
Bibi Maryam 110
Bidbid 141, 142
Bimah 191, 193
Bimmah 88
Birkat al Mouz 199, 202, 204, 207
Black smoker 239, 240, 241
Blueschist 44, 47
Boudinage 83, 253, 254, 255, 256
Bronze Age 5, 11, 12, 13, 121, 163, 220,
    221

Cairn grave 11, 14, 163
Calcite precipitation 174, 175, 176, 203,
    230, 231, 232, 233
Cambrian 36, 154, 158, 159, 161
Carbon cycle 54
Carbon dioxide 54, 188

Carbonate platform 36, 37, 56, 161, 200, 203, 204, 206
Carbonatite 114, 127
Carboniferous 36, 44, 88, 116, 155
Cave 40, 92, 93, 160, 203
Chert 45
Chevron fold 118, 217
Chilled margin 225, 226, 227, 235, 236, 247
Chromite 57, 58, 127, 135, 179, 182
Clay mineral 59, 60
Climate 16, 17, 40, 54, 94, 131, 133, 168, 188
Coal 58, 59, 60
Coast 19
Colemans's rock 195, 196
Collision 41, 46
Columnar basalt 228
Conglomerate 107, 113, 124, 145, 157, 166, 172, 186, 187
Continental crust 51
Continental margin 36
Copper 12, 14, 57, 58, 221, 237, 239, 241, 247
Copper Age 11
Coral reef 107, 199, 200
Cretaceous 36, 37, 38, 39, 41, 42, 51, 58, 59, 63, 72, 77, 110, 116, 119, 130, 164, 178, 194, 204, 207, 208, 210, 221, 223, 242, 250
Cumulate 46, 50, 136, 178, 179, 180, 181
Cryogenium 187
Crystalline basement 34
Cyclone 16, 122
Cyclone Gonu 16, 19, 106
Cyclone Phet 16, 19
Cyclone Chapala 16

Dammam formation 92
Date palm 28
Dead Sea 41
Devonian 36
Desert 20, 28, 32, 131, 147
Diamictite 53, 56, 88, 171, 172, 187
Dike 46, 78, 129, 131, 248

Dike-in-dike structure 49, 225, 233, 236
Diopside 49
Dhofar 6, 9, 10, 16, 18, 25, 26, 29, 31, 32, 59, 64
Dhow 73
Dolerite 34, 46, 78, 225
Dolomite 56, 58, 59, 71, 73, 75, 184, 187, 190, 192, 193, 203
Dune 8, 18, 19, 22, 31, 32, 33, 40, 131, 132, 133, 147, 148, 150, 151, 155, 156, 157
Dunite 48, 50, 58, 71, 73, 135, 136, 149, 179

Earthquake 102, 110
Eclogite 44, 47, 76, 77, 78, 81
Ediacara 158, 161
Eocene 41, 42, 73, 88, 91, 92, 95, 98, 99, 101, 104, 108, 110, 145
Escarpment 157
Evaporate 36, 151, 160

Fahud 65, 66, 206
Falaj 13, 14, 23, 165, 173, 197, 198, 203, 204, 207
Fiq formation 56, 187
Fiqa formation 59, 63
Fitri formation 208
Fluvial deposit 58, 91, 155, 157
Fold 74, 117, 204, 207, 249, 251, 252, 253, 258
Fort Bait al Rudaidah 204, 207

Gabbro 30, 45, 49, 51, 127, 178, 179, 181, 224
Garnet 48, 77, 78, 128
Gas 68
Ghaba Salt Basin 36, 160, 162
Gharif formation 36
Glaciation 52, 56, 188
Gold 57, 58
Gondwana 36, 116, 154, 159
Gossan 58, 59, 239, 240, 241, 243
Granite 129
Greenhouse 54, 188

Gubrah bowl 165, 169, 170, 171, 173, 194
Gulf of Aden 40, 149
Gulf of Oman 18, 44
Guyot 37, 220
Gypsum 58, 59, 151, 161

Habshan formation 204
Hadash formation 56, 187
Hafit tomb 11, 12, 97, 163, 220
Haima supergroup 63, 154
Hajar formation 63, 192, 193
Halban 162
Half-dike 49, 225, 236
Halophytic plants 19
Hamrat Duru basin 37, 38, 251, 253, 258
Hamrat Duru group 38, 204
Hamriyah 73
Harzburgite 49, 58, 71, 73, 135, 136, 164, 179, 257
Hasat Bin Salt, 13, 195, 196
Hatat formation 36, 85
Haushi group 63
Hawasina 30, 59, 131, 220, 230, 257
Hawasina basin 37, 38, 44, 52, 179, 216, 250
Hawasina complex 44, 220
Hawasina group 249, 250, 251, 252, 253, 256, 257, 258
Hawasina nappe 44, 138, 143, 170, 207, 216, 217, 218, 220, 223, 250, 257
Hawiyat Najm Parc 88
Haybi 37, 44
Hayl al Ashqarain 223
High-pressure metamorphism 46
Holocene 9, 17, 133, 214
Homo erectus 7
Homo sapiens 7, 8, 9
Horn of Africa 8, 11
Hot spring 31, 164, 176
Huqf 6, 32, 36, 63, 132, 187
Hydrocarbon 60, 61, 62, 67, 68
Hydrothermal vein 85
Hydrothermal fluid 239, 240, 244, 245
Hyaloclastite 46, 247

Ibri 59
Ice Age 33
Interglacial period 17
Iron Age 13, 14, 83
Islamic period 15
Island arc 58
ITCZ, Intertropical Convergence Zone 16, 17

Jafnayn formation 76, 110
Jalan Bani Buhassan 34, 127, 129
Jebel Akhdar 31, 35, 36, 37, 39, 56, 187, 192, 204, 208
Jebel Al Kamar 32
Jebel al Qal'ah 195
Jebel Aswad 225
Jebel Ja'alan 132
Jebel Khadar 92
Jebel Khamis 123
Jebel Kawr 215
Jebel Misht 219, 220, 221
Jebel Moqalit 225
Jebel Qahwan 110, 130
Jebel Qara 6, 9, 10
Jebel Samhan 32
Jebel Shams 23, 25, 30, 211, 220
Jiddat al Harasis 18
Jurassic 63, 116, 119, 204, 206, 209, 210, 218, 250, 257

Kahmah group 208, 211, 212, 223
Kaolinite 59
Karstification 31, 40, 88, 92, 98, 103, 104, 138, 173, 202, 203, 214
Kashlat Meqandili see Majlis Al Jin
Khareef 16, 25
Kharus formation 184
Khuff formation 36, 155
Kimberlite 127

Lake deposit 168, 214
Land use 28
Landslide 165
Lapilli 128
Lava tube 226, 227

Lekhwair 206
Lherzolite 48, 49
Limestone 57, 59, 74, 75, 82, 91, 95, 99, 104, 108, 138, 157, 164, 165, 184, 190, 203, 208, 211, 223
Listwanite 143, 144
Lithospheric mantle 134
Lizq 14
Lusail 237

Magan, country of 12, 58, 221, 237
Magma chamber 49, 58
Magnesite 69, 145, 183, 233
Mahil formation 72, 75, 200, 207
Majlis Al Jin 31, 92, 93
Makganyene glaciation 52, 56
Makran Subduction Zone 34, 42, 102
Manganese 57, 59, 119, 244
Mangrove 19, 76, 111
Marble 59, 76, 77
Marinoan glaciation 52, 57
Masirah 32, 151
Masirah Ophiolite 32, 149
Megalodont 212, 213
Mesolithic 9
Mesozoic 87, 204
Metamorphic rock 44, 75, 76, 78, 85, 142, 172
Metamorphic sole 51, 141, 179, 249, 250
Methane 54, 55
Mica schist 76, 77, 143
Mid-oceanic ridge 47, 49, 142, 224
Mineral resources 57
Miocene 42, 113, 157
Miqrat 153
Miqrat formation 155
MIS, Marine Isotope Stage 9
Misfah formation 213, 216
Miskin 257
Mistal formation 36, 171, 187
Moho, Mohorovičić discontinuity 30, 50, 58, 134, 136, 149, 176, 179
Monsoon 17, 25
Monsoon, Indian Ocean 17, 133
Muaydin formation 190

Mud flat 19
Muhammad 15
Mullion 253, 254, 255, 256
Musandam 30, 31
Musandam mountains 18, 31
Muscat 39, 59, 69, 73, 162
Muti formation 72
Mylonite 51

Nabkha 20
Nafun formation 36
Nakhl 31, 164, 183
Nakr Umr formation 209, 210, 212, 223
Natih 65, 204
Natih formation 63, 204, 208, 209, 210, 212, 223
Nejd plateau 9
Neolithic 6, 10, 11, 92, 126, 195
Neotethys 34, 36, 37, 38, 39, 41, 42, 44, 45, 51, 59, 116, 139, 178, 200, 204, 216
Nile 8
Nizwa 13, 59, 165, 197, 198, 204
Non-conformity 73, 91, 170, 173, 184, 185, 200
Notch 113
Nummulite 92, 145

Obduction 37, 41, 43, 44, 45, 46, 47
Ocean floor 47, 226
Oceanic crust 30, 49, 134
Oceanic lithosphere 45, 47, 48
Ocher 246
Oil field 65, 66, 67, 160
Oil see hydrocarbon
Oligocene 42
Olivine 48, 50, 181, 233
Oman exotic 37, 59, 137, 139, 216, 219
Opal 246
Ophiolite 30, 44, 45, 46, 50, 51, 52, 58, 174
Ordovician 36, 83, 85, 87
Owen Fracture Zone 34, 40, 109, 110
Oyster 74, 100, 102, 124, 125

Palaeolithic 5, 6, 9, 157
Palaeozoic 36, 62, 86, 87, 154, 160
Pangaea 184, 187, 200
Partial melting 47, 48
Parthian empire 14
Pegmatite 129, 131
Pelagic limestone 46
Pencil cleavage 86, 172, 190
Peridotite 45, 48, 51, 72, 144, 178, 179, 181, 233
Permian 36, 37, 44, 73, 74, 75, 80, 116, 138, 155, 165, 170, 173, 184, 199, 200, 203, 250
Persian empire 14
Phengite 77
Phyllite 85
Pillow lava 30, 45, 49, 173, 224, 233, 234, 247, 248
Plagioclase 49, 54, 181, 235
Plagiogranite 46, 139, 178
Playa 32
Pleistocene 6, 17, 91, 97, 113, 157, 158, 169
Pliocene 41
Pluvial 17
Polje 214, 215
Porcellanite 117
Precambrian 36, 62, 129, 159, 170, 171, 173, 184, 194
Proterozoic 36, 52, 53, 55, 56, 57, 83, 85, 130, 158, 173, 187, 188, 189, 190, 192
Pyramid see ziggurat
Pyroxene 48, 50, 54, 77, 181, 233, 235

Qahlah formation 110
Qalhat 109, 110
Qarat Kibrit 158
Quartzite 51, 85, 143
Quaternary 17, 40, 91, 99, 107, 125, 131, 133, 145, 170, 171
Quriyat 10, 108

Radiolarite 46, 116, 120, 127, 220, 251, 253

Rainfall 16
Ramlat Fasad 10
Ras al Hadd 10, 14, 30, 114, 121
Ras al Hamra 10
Ras al Jinz 10, 126
Red Sea 8, 30, 40, 149
Rhyolite 138
Ripple marks 83, 151
Rodinia 53, 188
Rub al Khali 18, 21, 33
Rudist 165, 208, 210
Rustaq 165

Sabkha 20, 21, 32, 123, 151, 155
Sahban spring 230
Sahtan bowl 192, 194
Sahtan group 209, 211
Saih Hatat 31, 36, 39, 42, 44, 76, 78, 80, 85
Saiq formation 74, 75, 77, 187, 200
Saiq plateau 25, 173, 199, 200, 201, 202, 203, 204
Saiwan 6, 156, 157
Salalah 29
Salt 123, 151, 153, 158, 160
Salt diapir 158, 159, 160
Samad period 14
Samail Ophiolite 11, 30, 31, 32, 38, 39, 42, 43, 44, 46, 47, 50, 52, 58, 69, 71, 104, 134, 136, 164, 178, 223, 226, 228, 229, 233, 234, 240, 242, 245, 249, 257
Samhan formation 60
Sandstone 153
Sasanian empire 14
Seamount 139
Seeb formation 145
Selma plateau 92, 95, 97
Serpentinite 45, 106, 136, 182, 183
Shamal 16, 31
Sharijah 203
Sharqiyah 6, 32, 33
Sheath fold 78
Sheeted dike complex 48, 223, 224, 229

Shell midden 10, 14, 125
Shenna 151
Shinas 58
Shuaibah formation 204
Sid'r formation 222
Sink hole 88, 92, 93, 98, 203
Sint 212, 214, 215
Snake Gorge 191, 192
Snowball Earth 52, 53, 54, 55, 56, 57, 187, 188
Sohar 13, 14, 58, 223, 233, 240, 242
Spilitization 226
Spreading center 47, 48, 49, 51, 58, 178
Star dune 147, 156
Stone Age 6
Stromatolite 184, 186, 189
Sturtian glaciation 52, 55, 56, 187
Subduction 37, 42, 46, 47, 51
Sumeini group 37, 44
Sur 10, 59, 60, 108, 109, 111, 112, 114
Sur formation 113
Suture zone 46, 47
Suwayh 10

Tawiyah 182
Terrace 97, 108, 164, 202
Tertiary 42, 44, 59, 60, 71, 148
Tholeiite 50
Thrust fault 193
Tillite 53, 172, 187
Tiwi 10, 107
Tower tomb 95, 162
Travertine 32, 162, 202, 231
Triassic 36, 37, 41, 51, 71, 75, 81, 138, 200, 204, 213, 214, 216, 218, 250, 257
Tsunami 99, 100, 101, 102, 122
Tufa 202, 203

Ultramafic rock 69, 71, 104, 134, 135, 143, 144, 182, 183, 230
Umar basin see Hamrat Duru basin
Umm an Naar tomb 12
Umm as Samim 20, 32

Vegetation 18
VMS, Volcanic hosted massive sulfides-deposits 237, 239, 240, 241, 242, 245
Volcanic glass 46, 235
Volcanic rock 148

Wadi ad Dil 249, 250
Wadi al Abyad 174, 176
Wadi al Alá 215, 216
Wadi al Ayn 217, 219
Wadi al Hat 193
Wadi al Hibi 223, 224, 226, 228
Wadi al Jizzi 58, 230, 231, 233, 234, 235, 236, 237
Wadi al Kabir 71, 257, 258
Wadi al Mayh 78, 80, 81, 82
Wadi al Nakhr 208, 209, 210, 211
Wadi Amdeh 83
Wadi Andam 32, 131
Wadi Bani Awf 189, 190, 191, 192, 194
Wadi Bani Khalid 28, 31
Wadi Bani Kharus 184, 194
Wadi Batha 32, 131
Wadi Daiqa 86
Wadi Dawkah 32
Wadi Fanja 142
Wadi Fins 91
Wadi Fisaw 60
Wadi Ghul 23, 196, 208
Wadi Haslan 56, 187, 190, 191
Wadi Hawasina 37, 248, 250, 256
Wadi Huw 73
Wadi Mistal 165, 168, 169
Wadi Msawi 60
Wadi Muaydin 202, 203, 204, 205, 206
Wadi Murri 222
Wadi Musfa 138
Wadi Qatif 14
Wadi Raisah 111
Wadi Salmiyah 113
Wadi Shab 10, 103
Wadi Shakalah 123
Wadi Simayh 228, 229
Wadi Suq 13
Wadi Tiwi 28, 31, 104

Wahiba Sand 18, 21, 32, 131, 133, 151
Wahrah formation 116, 118, 119, 129
Wakan 173
Wasia group 211, 212
Weathering 54, 203, 214
Wehrlite 181
Westerlies 16
Whie smoker 245

Xenolith 127, 128, 140, 141, 148, 149

Yankul 14, 223
Yemen 16, 31
Yenkit 73
Yibal 65, 206
Yiti 71, 73, 76

Zagros fold-and-thrust belt 40
Ziggurat 237

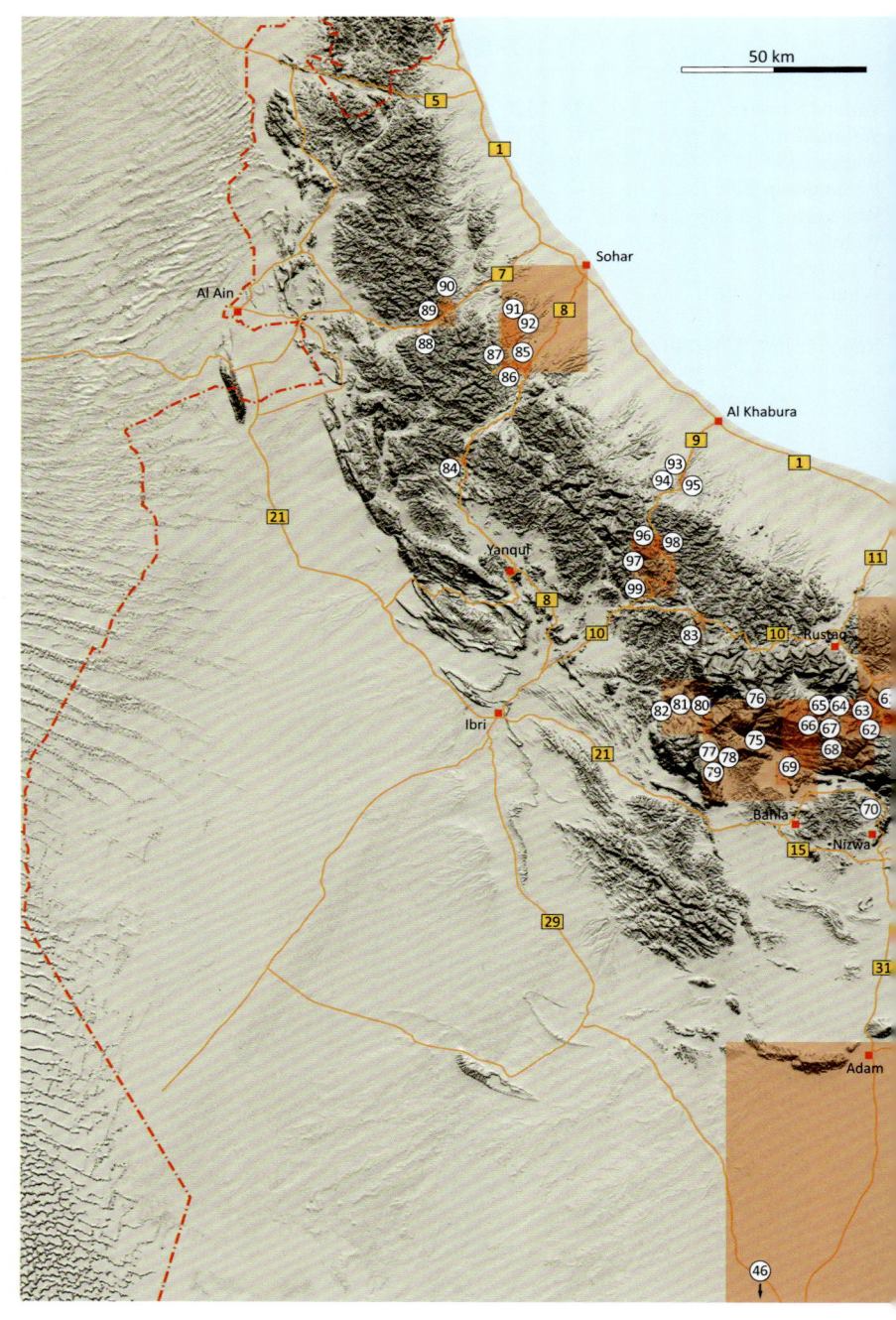

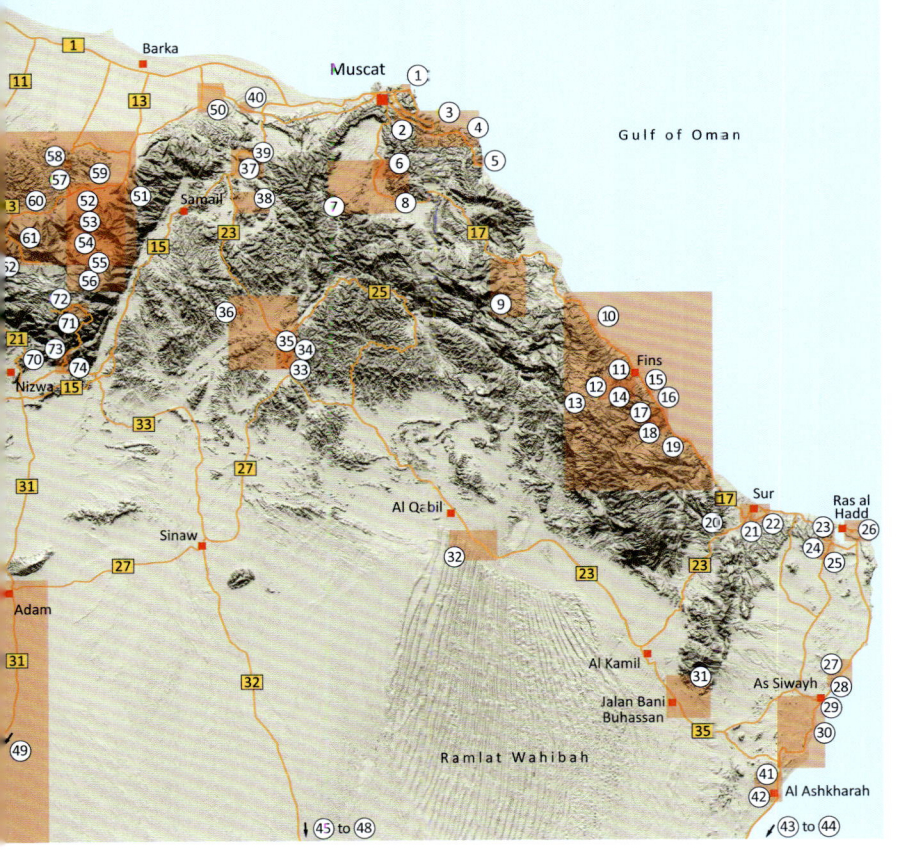

Gabi Schneider

The Roadside Geology of Namibia

(Geological Field Guides – Sammlung geologischer Führer, vol. 97)

2008. 2. edition, IX, 294 pages,
112 figures, 1 table, 14 x 20 cm

ISBN 978-3-443-15084-6,
paperback € 38.–

www.borntraeger-cramer.com/ 9783443150846

Namibia has over the years attracted scientists from all over the world to study its geology, uniquely exposed in the desert environment. Their research has shaped geological thinking worldwide, and led to the development of many new concepts.

Due to an arid climate and low population density, geological features are ever present and eye-catching in Namibia. It is for these reasons, that both scientists and laymen are attracted to the country, and many a tourist develops a keen interest in geology when touring this beautiful country.

The Roadside Geology Guide has been prepared on response to a growing demand for geological information. The guide provides a general introduction into the geological development of southern Africa and Namibia in particular, followed by general chapters on mineralogy, palaeontology, hydrogeology and mining. Special geological attractions of Namibia are then described in detail, followed by detailed route descriptions of 29 routes throughout the entire country. They are presented in a radial fashion, starting with routes from the central part of the country towards the west, followed by southern and eastern Namibia and ending with the northern routes. The route descriptions are accompanied by geological maps as well as stratigraphic diagrams.

In this second, revised edition, text, figures and cover have been corrected and optimized.

 **Schweizerbart Borntraeger**

Johannesstr. 3A, 70176 Stuttgart, Germany.
Tel. +49 (711) 351456-0 Fax. +49 (711) 351456-99
order@borntraeger-cramer.de   www.borntraeger-cramer.de

Patrick Stäheli:
Kalifornien I
Süden und Osten

Basin und Range, Transverse und Peninsular Ranges, Death Valley, Mojave-Wüste, Geologie und Exkursionen

[Field Guide to the Geology of California Part I: Southern and Eastern California]

(Sammlung geologischer Führer, Band 108)

2013. X, 275 Seiten, 214 überwiegend farbige Abbildungen,
14 x 20 cm,
ISBN 978-3-443-15096-9,
brosch., 29.90 €

www.borntraeger-cramer.com/ 9783443150969

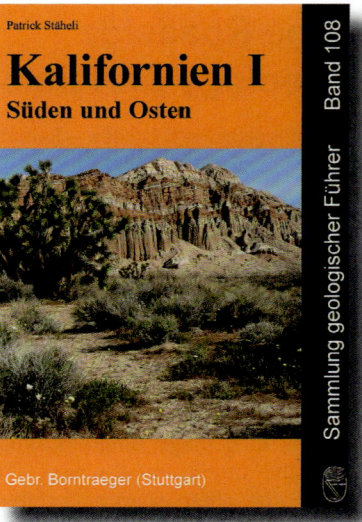

Der Band beschreibt die erdgeschichtliche Entwicklung Süd- und Ostkaliforniens. Nach einem allgemeinen Überblick über die geologische Struktur und Geschichte der Region werden mehr als achtzig landschaftlich reizvolle Reiseziele an der Küste und in den Wüsten, zu Vulkanen und Erdbebenverwerfungen sowie zu Sehenswürdigkeiten in den größeren Städten vorgestellt und deren geologische Bedeutung und Besonderheiten beschrieben. Anhand zahlreicher farbiger Illustrationen wird die Geologie dieser Reiseziele anschaulich und detailliert erklärt.

Die bekannte San Andreas Störung, die mit regelmäßigen Erdbeben die Angst vor dem nächsten „Großen Beben" in Kalifornien wach hält, wird ebenso erläutert wie die Geologie des Death Valley National Parks, der Long Valley Caldera bei Mammoth Lakes, des Anza Borrego Desert State Parks oder des Joshua Tree National Parks.

Auch viele weniger bekannte, aber geologisch ebenso interessante Punkte, wie die Fossil Falls oder der Painted Canyon, werden hier behandelt. Wegbeschreibungen mit GPS-Koordinaten und Detailkarten erleichtern die Routenplanung, ein Orts- und Stichwortverzeichnis zum raschen Auffinden relevanter Informationen rundet den Band ab.

Der Band richtet sich an alle, die sich für die geologische Geschichte Kaliforniens interessieren; er sollte bei keiner Exkursions- bzw. Reiseplanung fehlen.

Schweizerbart   Borntraeger

Johannesstr. 3A, 70176 Stuttgart, Germany.
Tel. +49 (711) 351456-0 Fax. +49 (711) 351456-99
order@borntraeger-cramer.de   www.borntraeger-cramer.de

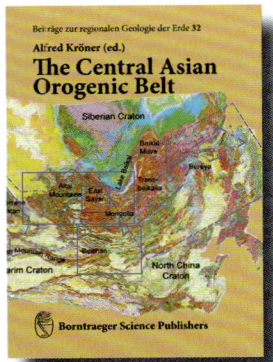

The Central Asian Orogenic Belt
Geology, Evolution, Tectonics, and Models

(Beiträge zur regionalen Geologie der Erde, Band 32)

Ed.: Alfred Kröner
2015. 313 pages, 109 figures, 2 tables, 18 x 25 cm,
ISBN 978-3-443-11033-8, bound, 118.00 €

www.borntraeger-cramer.com/ 9783443110338

This volume provides a state-of-the-art account of the geology of part of Central Asia named The Central Asian Orogenic Belt (CAOB). This Belt formed by accretion of island arcs, ophiolites, oceanic islands, seamounts, accretionary wedges, oceanic plateaux and microcontinents (c. 1000–250 Ma ago) by similar processes to those in the circum-Pacific Mesozoic–Cenozoic accretionary orogens. Also known as Altaids, this region is one of the largest orogenic belts on Earth, extending from the Ural Mountains in the West to far eastern Siberia.

In view of the increasing significance of Central Asia because of its wealth of mineral resources this volume is of interest to readers from all fields of the geosciences and from academics to industry.

Planet Earth - In Deep Time
Palaeozoic Series
Devonian & Carboniferous

Ed.: Thomas J. Suttner; Erika Kido; Peter Königshof; Johnny A. Waters; Laura Davis; Fritz Messner

2016. 261 pages,
201 coloured figures,
30 x 22 cm, ISBN 978-3-510-65335-5,
bound, 49.90 €

www.schweizerbart.de/9783510653355

Devonian and Carboniferous deposits are found in many places around the world.
The record of climate change preserved in them is the best (and only) resource of information on the Earth`s climate system then, at a time, when current and future climatic change is one of society`s greatest challenges.
This book introduces some of the key areas of Mid-Palaeozoic sediment occurrence worldwide, authored by 114 specialists from more than 30 countries. The areas were studied as part of the UNESCO/IUGS project on climate change and biodiversity patterns in the mid-Palaeozoic (Devonian and Carboniferous).
This large-scale taxonomic, stratigraphic and palaeoecological study of mid-Paleozoic floras and faunas has resulted in 86 contributions and more than 25 artistic reconstructions characterizing the biosphere of the Devonian and Carboniferous.

Geological Field Guides
Sammlung geologischer Führer – available volumes

Band 108: **Kalifornien I** [Süden und Osten]
Band 107: **Kreta**
Band 106: **Spessart** [Geologische Entwicklung und Struktur, Gesteine und Minerale]
Band 105: **Die deutsche Ostseeküste**
Band 104: **Harz, östlicher Teil mit Kyffhäuser Kristallin**
Band 103: **Karlsruhe und seine Region** [Nordschwarzwald, Kraichgau, Neckartal, Oberrhein-Graben, Pfälzerwald und westliche Schwäbische Alb]
Band 102: **Der Schwarzwald und seine Umgebung**
Band 101: **Aachen und nördliche Umgebung** [Mechernicher Voreifel, Aachen-Südlimburger Hügelland und westliche Niederrheinische Bucht]
Band 100: **Aachen und südliche Umgebung** [Nordeifel und Nordost-Ardennen]
Band 99: **Korsika** [Geologie, Natur und Landschaft, Exkursionen]
Band 98: **Elba**
Band 97: **The Roadside Geology of Namibia**
Band 96: **Nördliches Harzvorland**
Band 95: **Thüringer Wald**
Band 94: **Die Hochrhein-Regionen zwischen Bodensee und Basel**
Band 93: **Wetterau und Mainebene**
Band 92: **Das Ries und sein Vorland**
Band 91: **Ungarn**
Band 90: **Nordwürttemberg**
Band 89: **Das Rheintal zwischen Bingen und Bonn**
Band 87: **Dresden und Umgebung**
Band 86: **Die Südalpen zwischen Gardasee und Friaul**
Band 85: **Thüringer Becken**
Band 84: **Saarland**
Band 83: **Italienische Vulkangebiete V**
Band 81: **Kanarische Inseln**
Band 74: **Mainfranken und Rhön**
Band 73: **Salzburger Kalkalpen**
Band 70: **Harzvorland – westlicher Teil**
Band 69: **Italienische Vulkan-Gebiete III**; [Lipari, Vulcano, Stromboli, Tyrrhenisches Meer]
Band 68: **Oberbergisches Land**
Band 67: **Die Schwäbische Alb und ihr Vorland**
Band 66: **Zwischen Jadebusen und Unterelbe**
Band 64: **Die Insel Elba und die kleineren Inseln des Toskanischen Archipels**
Band 62: **Hegau und westlicher Bodensee**
Band 59: **Wienerwald**
Band 58: **Harz. Westlicher Teil**
Band 57: **Das ostfriesische Küstengebiet**
Band 55: **Ruhrgebiet und Bergisches Land**
Band 53: **Ötztaler und Stubaier Alpen**
Band 50: **Fränkische Schweiz und Vorland**
Band 49: **Vorarlberger Alpen**
Band 48: **Aachen und Umgebung**
Band 42: **Das Steirische Randgebirge**

More information

www.borntraeger-cramer.com/series/sgeolf